Visions of a Vanished World

VISIONS OF A VANISHED WORLD

The Extraordinary Fossils of the Hunsrück Slate

Gabriele Kühl,
Christoph Bartels,
Derek E. G. Briggs,
and Jes Rust

Foreword by Richard Fortey

Yale UNIVERSITY PRESS
New Haven and London

Originally published as *Fossilien im Hunsrück-Schiefer: Einzigartige Funde aus einer einzigartigen Region,* by Gabriele Kühl, Christoph Bartels, Derek E. G. Briggs, and Jes Rust, copyright © 2012 by Quelle & Meyer Verlag GmbH & Co., Wiebelsheim. Translated and revised by the authors.

Yale University Press books may be purchased in quantity for educational, business, or promotional use. For information, please e-mail sales.press@yale.edu (U.S. office) or sales@yaleup.co.uk (U.K. office).

Set in The Sans and The Serif type by BW&A Books, Inc. Printed in China.

Library of Congress Cataloging-in-Publication Data
Fossilien im Hunsrück-Schiefer. English
Visions of a vanished world : the extraordinary fossils of the Hunsrück Slate / Gabriele Kühl, Christoph Bartels, Derek E. G. Briggs, and Jes Rust ; foreword by Richard Fortey.
  p. cm.
  Includes bibliographical references and index.
  ISBN 978-0-300-18460-0 (hardback)
  1. Paleontology—Devonian. 2. Marine animals, Fossil—Germany—Hunsrück. 3. Plants, Fossil—Germany—Hunsrück. 4. Hunsrück Shale (Germany)   I. Kuhl, Gabriele, 1975–   II. Title.
  QE728.F6713 2012
  560'.1740943431—dc23   2012013826

A catalogue record for this book is available from the British Library.

This paper meets the requirements of ANSI/NISO Z39.48–1992 (Permanence of Paper).

10 9 8 7 6 5 4 3 2 1

Contents

Foreword

Thirty years ago I was visited by a German professor in my office in what was then known as the British Museum (Natural History) in London. He had with him an astonishing folio of photographs; they were x-radiographs of fossils that were 400 million years old, revealing many intimate details, painted in ghostly grays, of the jointed legs and antennae of long extinct arthropods. The professor was Wilhelm Stürmer, and the fossils were from the Hunsrück Slate, a Devonian-age formation exposed south of Koblenz. The profound impression those fossils made has never left me. The Hunsrück is one of the greatest examples of a Konservat-Lagerstätte—those incredibly rare occurrences where nearly all the inhabitants of an ancient seafloor are preserved complete, to give us a vision of a vanished marine world in all its complexity. The fossil record usually affords us no more than a glimpse of the whole habitat: just those organisms favored with shells that survive the vicissitudes of fossilization. But here it is more like traveling back in a time machine to see a seafloor almost as in a documentary film, as one marvel after another is revealed. Stürmer's x-rays picked up hidden details, showing what lay concealed in the rock.

Now several decades have passed, and the persistence and skill of preparators has allowed many of these details to be exposed to view directly, as the patient techniques for extracting fossils from the rock have been refined and new technology has been developed. What was seen "as in a glass, darkly" can now be seen in clear view. Gabriele Kühl and her colleagues have brought together a magnificent collection of photographs of these extraordinary fossils. The "sea lilies" seem to have been preserved in motion, still waving their arms in the ancient sea in pursuit of fine food particles. Brittle stars crawl in profusion over the mud, now hardened to black plates. Trilobites peer out at their contemporaries through their crystalline eyes. The cliché "frozen in time" for once is absolutely right—the Hunsrück Slate is a comparatively short moment of geological time where decay has been cheated, and it is possible to join strange and wonderful animals leading their varied lives not long after the very first animals and plants had moved on to the land, and when dinosaurs were no more than a twinkle in the evolutionary eye. It is indeed a strange world, but also, paradoxically, a partly familiar one. Many of the animals, and particularly the arthropods, are not like their distant relatives that survive today. Like the more familiar trilobites, they belong to vanished families, long extinct. They help us to understand how the modern biological world came about, yet they have had their moment thronging the seafloor, and have been replaced by other creatures. But there are also fossils of what are undeniably sea spiders, whose relatives still can be found in today's oceans, as survivors from ancient times. The clots of brittle stars, too, look at a glance like recent documentary footage of life on the floor of the ocean abyss taken by high-tech underwater cameras. So the Hunsrück provides a fascinating mixture of the bizarre, the unknown, and the half familiar. The authors take us on a tour through all the different animal phyla, explaining the significance of the fossils to our understanding of the biosphere. To a paleontologist, these pictures are not merely scientifically fascinating, they are also aesthetically pleasing. An artist would enjoy the feathered brushes of the sea lily arms or the filigree hairs on the legs of *Mimetaster*. There is every reason to suppose that the seafloors of the past were almost as diverse and richly used for their resources as those of the present day. With this book we may take a kind of mental bathyscape down to the deeps of the Paleozoic Era, probing the seafloor like a marine biologist given the chance to transcend time and space. It is a world well worth exploring.

Richard Fortey, FRS FRSL

Preface

"Das Schöne zieht einen Teil seines Zaubers aus der Vergänglichkeit." [Beauty owes some of its charm to its transitory nature.]—Hermann Hesse

Scarcely anything is more impressive and yet more transient than the diversity of life. But not every trace of past life is extinguished; fossils provide a window on the organisms and environments of millions of years ago. The Hunsrück region of Germany has yielded some of the world's most extraordinary fossils. The spectacular appearance and often complete preservation of the specimens are as remarkable as the uniqueness of many of the organisms. The outstanding quality and quantity of fossils from the Hunsrück region have made it internationally famous. And the Hunsrück Slate is the world's most important marine fossil deposit of Devonian age (416 to 359 million years ago). As such, it is an extraordinary window into the history of life on Earth and offers unique insights into the evolution of marine biodiversity.

In this book we introduce readers to the Hunsrück Slate, the diversity of organisms and their modes of life, and, above all, the beauty of the fossils. Although all the world's major natural history museums hold examples of Hunsrück Slate fossils, it is impossible to create a list of all the known specimens. Nevertheless, we have used some of the most spectacular examples to illustrate life in the Devonian seas.

Acknowledgments

We are grateful to everyone involved in the completion of this volume. The excellent work of photographers Georg Oleschinski and Alexandra Bergmann made this book possible. Herbert Lutz of the Naturhistorisches Museum Mainz/Landessammlung für Naturkunde Rheinland-Pfalz and Michael Wuttke of the Direktion Landesarchäologie, Generaldirektion Kulturelles Erbe RLP, Mainz, allowed us to photograph and study fossils in their care. The Deutsches Bergbau-Museum (German Mining Museum), Bochum, provided access to fossils from the Bartels collection and other material, as did the Goldfuß Museum of the Steinmann Institute of the University of Bonn. We thank the Alexander von Humboldt-Stiftung and the Deutsche Forschungsgemeinschaft for funding the collaboration that led to the emergence of the idea for the book and the involvement of the authors. Jean Thomson Black, our editor at Yale University Press, and Erica Champion provided advice and assistance in the preparation of this English version of our original German text. Derek Siveter and Paul Selden reviewed an earlier version of the English manuscript for Yale University Press and offered valuable comments.

THE FOSSILS OF THE HUNSRÜCK SLATE

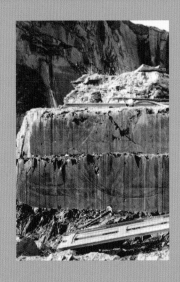

Cenozoic	Quaternary	2.6
	Tertiary	65
Mesozoic	Cretaceous	145
	Jurassic	200
	Triassic	251
Paleozoic	Permian	299
	Carboniferous	359
	Devonian	416
	Silurian	444
	Ordovician	488
	Cambrian	542
	Precambrian	

Hunsruck Slate (at Devonian, 416)

Figure 1. The position of the Hunsrück Slate in the geological time scale. The numbers at right are millions of years before the present.

The Hunsrück region of Germany is a scenic area of low mountains with forests and meadows west of the river Rhine, bounded by the rivers Moselle, Nahe, and Saar. It lies in the middle of the state of Rhineland-Palatinate and includes a small part of the Saarland. Geologically, the Hunsrück region consists mainly of Lower Devonian slates, sandstones, and quartzites, which date to about 400 million years ago (Figure 1).

The Hunsrück Slate is named after the region but extends beyond it to the north and east; the slate is exposed at the surface in the mountainous area between Mayen in the north and Bingen in the south and reaches a total thickness of about 3,750 meters (Figure 2). Fossils are present throughout, but finding them often requires extensive searching. However, evidence of the soft-part anatomy of the animals is preserved only in the slate quarries near the towns of Bundenbach, Gemünden, and Herrstein. Superb fossils without soft parts have also been found in other regions where the Hunsrück Slate is exposed, such as near Altlay (Hünsruck), in the Wisper River valley (Taunus), and in the area around Mayen (southeastern Eifel). But fossils are much rarer in these areas (where research is ongoing) than in the central Hunsrück region.

The sequence preserving the best fossils consists of about 150 meters of mostly uniform mud with rare layers of fine sand. The fossils were buried in layers of sediment on the seafloor. The surfaces of these layers or beds are referred to as bedding planes. Mountain building in this region during the Devonian and Carboniferous resulted in subsidence, compression, folding of beds, and heating of the rocks to about 400 degrees Celsius. Even the shelly fossils are typically flattened. These tectonic forces also reoriented the clay minerals in the mud until they lay at right angles to the direction of compression, resulting in the formation of a planar fabric or cleavage in the rock. In this way the shales were converted to slate, which splits along the cleavage planes. The best chance of recovering intact fossils occurs where the slate splits parallel to the sedimentary layers. In contrast, where the slate splits at a high angle to these layers (that is, cleavage is at a high angle to bedding) fossils that have survived for millions of years are sliced up and destroyed.

Exposures of the Hunsrück Slate were known as early as Roman times. The slate was mined and used on roofs and the outer walls of buildings. For the first sixty years of the twentieth century, the Hunsrück Slate was extensively exploited by the mining industry, but the use of slate as a building material declined following the development of synthetic materials (such as "artificial slate") in the late 1950s. The oil crisis of the 1970s increased the price of synthetics and made slate mining in the Hunsrück region profitable again. Most of these mining operations ended in 1999, however, as the best and most accessible slate had already been mined or extracted, and slate could be imported more cheaply from Spain. Today only three pits are still in operation, two at Mayen and one in the Altlay region near the northern edge of the Moselle Valley.

As is often the case in paleontology, economic interest in the rocks led to the discovery of the fossils that were hidden there. The scientific significance of the Hunsrück Slate fossils was recognized only in the second half of the nineteenth century. Scientists and fossil collectors were not the only ones who searched in the slates. The vast majority of specimens were discovered by the workers in the quarries, who supplemented their meager wages with the sale of the fossils they found when splitting the roofing slate. The efforts of fossil collectors and scientists yielded much fewer fossils. Nonetheless the number of fossils accumulated in public and private collections has grown steadily since about 1850. Today the fossils of the Hunsrück Slate are among the most sought after, and they are represented in paleontological collections in Germany and in several major museums around the world.

Formation of the Hunsrück Slate

The distribution of the continents some 400 million years ago, when the Hunsrück sediments were deposited and the fossils formed, was entirely different from today. In the southern hemisphere the supercontinent Gondwana formed a huge land mass, while to the north of it lay the landmass Laurussia, often called the Old Red Continent. Gondwana and Laurussia were separated by the Rheic Ocean. The Hunsrück sea occupied a flat shelf area along the northeast to southwest trending coast of Laurussia during the Devonian.

The mud and sand deposits that formed the Hunsrück Slate were concentrated in the coastal areas of Laurussia and extended into the ocean basin. The water was less than 200 meters deep, and the sea experienced warm tropical temperatures because of its proximity to the equator.

According to Bonn geologists Johannes Stets and Andreas Schäfer, the seabed was divided by ridges that ran parallel to the ancient coastline.[1] Seismic activity and possibly severe weather (such as occurs in tropical latitudes today) resulted time and again in the transport of muddy sediments by turbidity currents into deeper areas of the basin. This action caused the sudden and rapid burial of the animals living in some areas (Figure 3). Accumulation of sediments, followed by tectonic processes, resulted in the formation of shale followed by folding and its subsequent transformation to slate. Uplift and the erosion and weathering of overlying deposits eventually exposed the deeply buried slate at the surface of the Hunsrück Mountains.

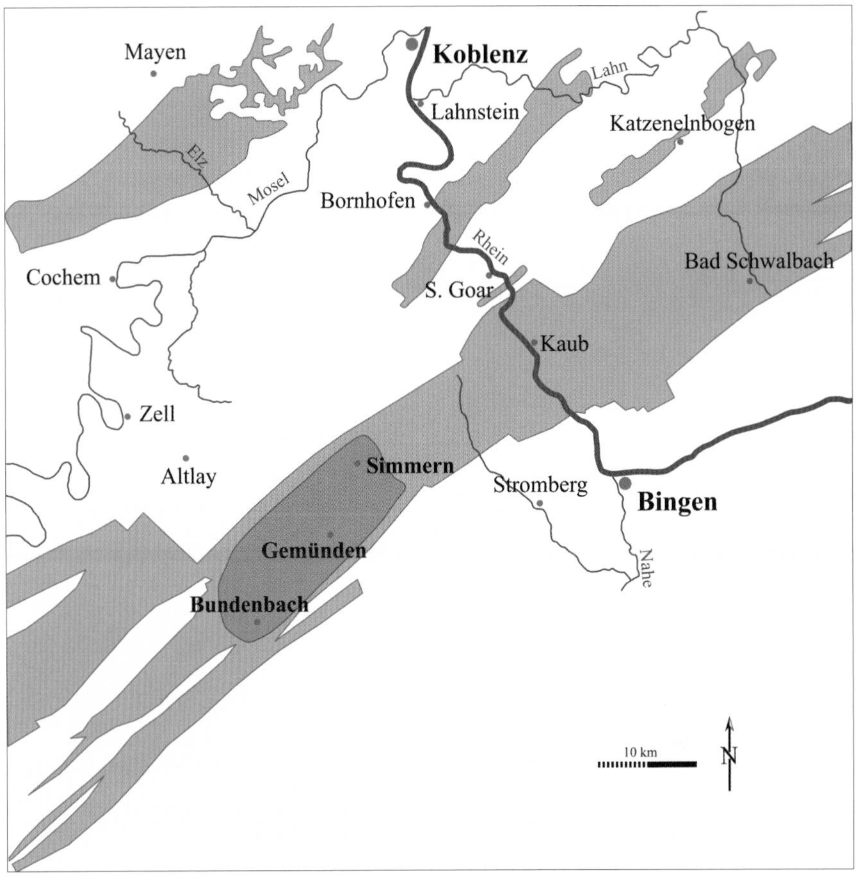

Figure 2. Simplified map of the distribution of the Hunsrück Slate in the Rhenish Slate Mountains. The famous fossils are from the area around Bundenbach, Gemünden, and Herrstein (about 6 kilometers southwest of Bundenbach).

Preservation of the Animals as Fossils

Only a small proportion of the marine organisms living in the Hunsrück sea were buried intact in the sediment. The vast majority of the many species of animals and the much less familiar plants went through the usual processes of decay and degradation, resulting in the breakdown of the organism. Several animals are represented only by hard parts such as shells and other skeletal elements. Only very rarely are skeletons preserved in their original arrangement. Most of the Hunsrück Slate fossils are the disarticulated remains of the hard parts of marine animals.

Rapid burial in mud does not automatically lead to fossilization. As a rule, microbes and scavengers on the seafloor decompose animal and plant tissues to simpler molecules that reenter nature's cycle. Most traces of former life are erased. Where more favorable circumstances occur, decay and degradation may be inhibited and a fossil may form. In the case of soft tissues such as muscle, skin, tendons, ligaments, and cartilage, fossilization occurs only in exceptional cases, because these tissues are decomposed quickly by microorganisms. Protection from destruction is also necessary to facilitate the preservation of hard parts. Rapid burial by clouds of iron-rich muddy sediment, transported by turbidity currents, provided the perfect environment for fossilization in the Hunsrück sea and explains the often remarkable preservation of traces of soft tissues.

But why do the Hunsrück Slate fossils appear golden? They were preserved by a process called pyritization.[2] The reaction of iron and sulfide in the sediment resulted in partial or complete replication of the soft tissues in the mineral pyrite. Laboratory experiments have shown that the initial stages of pyritization can take place in only 80 days.[3] The pyritization of soft-tissue structures ranges from a diffuse tracing of outlines to the replication of details such as muscle strands or the gut trace of an animal. Pyritization occurs where dissolved iron in the pore water within the sediment combines with sulfur compounds that are generated by sulfate-reducing microbes. As a result of pyritization, not only is the preservation of the smallest details of hard parts possible, but so is evidence of soft-tissue morphology. A favorable side effect of pyritization is that the fossils become opaque to x-rays, in contrast to the surrounding slate. For this reason, the study of fossils need not rely only on surface exposure of specimens—details hidden in the rock can be imaged. This phenomenon is an invaluable asset in guiding the preparation of Hunsrück fossils and in revealing features concealed in the rock.

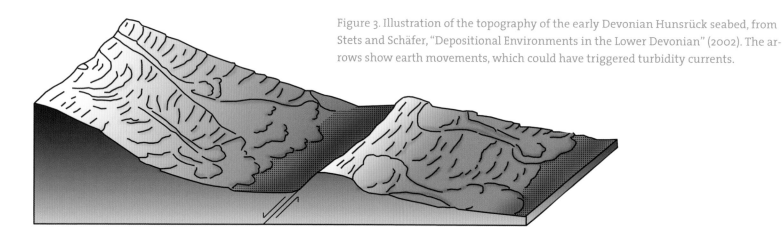

Figure 3. Illustration of the topography of the early Devonian Hunsrück seabed, from Stets and Schäfer, "Depositional Environments in the Lower Devonian" (2002). The arrows show earth movements, which could have triggered turbidity currents.

The Significance of the Hunsrück Slate Fossils

Detailed and complete preservation of fossils is extremely important for understanding past life forms. The quality of preservation of Hunsrück Slate fossils is unique in marine deposits of Devonian age. The fauna of the Hunsrück Slate is exceptionally diverse and includes more than 270 described species. The most common fossils are sea lilies, sea stars, brittle stars, and arthropods, such as trilobites and crustaceans. But sponges, corals, bryozoans, brachiopods, bivalves, cephalopods, sea cucumbers, and fish are also present. Many of the species are known only from the Hunsrück Slate. We can reconstruct the morphology of these animals, and interpret evidence of certain behaviors such as locomotion and feeding, to paint a picture of a bygone world.

The Hunsrück fossils not only record the time and place in which these animals lived, but they also provide important evidence about their evolutionary relationships with earlier animals, and life in the Devonian seas. Some of the fossils have a wider significance for our understanding of the evolution of life in the Paleozoic seas because they are closely related to Cambrian forms, some 100 million years older. A number of the arthropods, for example, prove that some groups characteristic of the Cambrian continued to flourish even in the Devonian.

Excavation and Preparation

Fossils are widespread in the Hunsrück Slate, but finding them requires considerable experience and a good eye. Spectacular discoveries have always been rare, and the relatively large number of specimens in some collections is the result of decades of effort. Most Hunsrück Slate fossils are much less well preserved than those pictured here. Specimens often show only a vague outline or are difficult to interpret due to deformation in the slate, making it a problem to identify them until they are prepared. Traditionally the splitters working in roof slate production found the fossils; they developed an eye for fossil treasures during their daily routine of working with the slate. Most institutions and private collectors acquired material turned up in this way. A handful of collectors explored the spoil heaps generated by slate production and discovered fossils themselves (Figure 4).

It requires a great deal of time and takes great patience, skill, and experience, as well as good preparation technique and equipment, to convert an unremarkable piece of slate into a beautiful, sought-after fossil specimen. If possible, a preparator will x-ray the slate before starting work to determine the exact outline and position of the specimen. Then the elaborate work of revealing the fossil can commence.

Many fossils were exposed by the quarry workers with mechanically rotating metal brushes. This method is quick, and the fossils acquire a thin brass coating with a "golden" sheen. A major disadvantage, however, is that fine structures and details are irretrievably lost, either through mechanical abrasion by the brush or a coating of metal debris. A more satisfactory method is the use of blades and needles. The preparator exposes the features of the fossil with finely sharpened tools while monitoring the process under a dissecting microscope. This method is very time- and labor-intensive and requires a steady, careful hand, but, if done well, it yields excellent results.

When appropriate equipment is available, finely sharpened scrapers can be used in combination with an air-abrasive machine that blasts the surface of the slate with fine iron powder. This method achieves excellent results, and the time required for preparation is significantly reduced. The equipment is noisy and expensive and requires appropriate space for its operation. With this method in particular the preparator needs to take precautions to avoid inhaling the silica-containing dust.

The air-abrasive method allows very delicate preparation and has the potential to retain all of the detail in the fossils. The nature of the abrasive is of crucial importance: the particles must be of appropriate size and neither too hard nor too soft. The preparator must be able to adjust the pressure of the stream of iron powder striking the slate. The specimen is enclosed in a sealed cabinet with a lid through which the fossil can be viewed, ideally with a microscope. A suction device captures the dust, and the iron powder can be recycled.

Figure 4. Christoph Bartels collecting on the spoil heap of the Eschenbach-Bocksberg quarry at Bundenbach.

Conflicting Scientific Interpretations

The Hunsrück Slate was valuable as building and construction material long before the fossils were discovered, although it cannot be exploited everywhere it is found; in many places the rock is buckled and does not split into uniform layers (Figure 5). The geologist Carl Ferdinand Roemer, who worked in Bonn and Breslau, was the first to study the fossils of the Hunsrück Slate from a scientific perspective. In 1842 he began to investigate the stratigraphy of the Rhenish Massif with the aid of fossils, and was the first to map rocks of early Devonian age in the region. He described starfish, brittle stars, and sea lilies from the roof slates.[4] Otto Follmann, a teacher at the Kaiser-Augusta High School in Koblenz, published the first monograph on Lower Devonian sea lilies (crinoids) in 1887, including examples from the Hunsrück Slate.[5]

Research on the fossils of the Hunsrück Slate developed primarily in two different directions. While paleontologists interested in taxonomy investigated the nature and systematic classification of the fossil organisms, others tried to reconstruct the sedimentary setting and the environment where the animals lived. The geologist and paleontologist Otto Jaekel argued in 1895 that the sediments were deposited under deep-sea conditions.[6] This view was later supported, in 1966, by the Tübingen paleontologists Adolf Seilacher and Christoph Hemleben.[7] Other scientists, such as Fritz Kutscher and Rudolf Richter, however, considered the environment to have been a shallow-shelf sea similar to the Wadden Sea, which lies off the coast of Denmark, Germany, and the Netherlands.[8] Christoph Bartels and Günther Brassel proposed in 1990 that the preservation of complete organisms, including evidence of their soft tissues, was the result of rapid burial that killed the animals.[9] In 1997 continuous blocks of Hunsrück Slate were sawn from the Eschenbach-Bocksberg slate quarry as part of an international research effort known as the Nahecaris Project.[10] We now know, through investigation of this material and other fieldwork, that the fossiliferous Hunsrück Slate was deposited by turbidity currents in an elongate, shallow marine basin. Geochemical studies have shown how turbidity currents can contribute to the conditions necessary for pyritization.[11]

Figure 5. Vertical layers of foliated slate in part of the abandoned Eschenbach-Bocksberg quarry at Bundenbach.

The analysis of drill cores has revealed the influence of tides in some areas and shown that sufficient oxygen and food were available to make the Hunsrück sea a favorable environment for marine life. The exceptionally preserved fossils were the result of relatively small scale short-term depositional events, and the affected areas were recolonized by marine animals very quickly.

The physicist Walther Lehmann was the first person to x-ray the fossils of the Hunsrück Slate on a systematic basis, opening up new possibilities for their investigation from 1932 onward.[12] Subsequently Wilhelm Stürmer used x-radiography to investigate the fossils between 1960 and 1986.[13] As head of the radiology department at Siemens Corporation, he was able to improve x-ray technology to a significant extent. He even designed an x-ray device that was installed in a Volkswagen bus, allowing the technique to be used to examine slate slabs in the field. Stürmer's field research was focused on the site of a large abandoned quarry, the Kaisergrube, at Gemünden, but he also x-rayed material from localities around Bundenbach. The x-ray techniques developed on the Hunsrück Slate fossils still influence paleontological methods today.

X-rays were an important basis for the work of many other paleontologists investigating pyritized fossils internationally. More recently, Hunsrück Slate fossils have been imaged using computer-aided tomography scanning, which generates a three-dimensional x-radiograph. The potential of this new method for interpreting Hunsrück Slate fossils has yet to be fully realized.

An Excavation in the Eschenbach-Bocksberg Slate Quarry at Bundenbach

Hunsrück Slate fossils were found mainly during the splitting of the rock for roof and wall slates. During the last period of slate quarrying the rock was extracted using explosives, excavators, and large-tracked

transport vehicles. This, together with the largely uniform nature of the sequence, allowed only the most rudimentary observations of the distribution of fossils through the thickness of the slate, and the nature of the sedimentology. Opportunities to relate the occurrence of fossils to variations in the sediments were even more limited during underground mining, which was an important element of slate production until 1960. Fossils and sedimentary structures are very difficult to observe in poorly lit underground galleries.

The fossil-bearing strata in the Eschenbach-Bocksberg slate quarry were systematically excavated in 1997, with the cooperation and support of the owners of the slate company Johann and Backes, in an area that was no longer actively exploited.[14] The project was led by Michael Wuttke, head of the Direktion Landesarchäologie, Generaldirektion Kulturelles Erbe RLP in Mainz (formerly the Landesamt für Denkmalpflege Rheinland Pfalz, Department of Geological Heritage), with collaboration from the University of Bristol in the United Kingdom (Derek Briggs) and the Deutsches Bergbau-Museum in Bochum, Germany (Christoph Bartels). An area of bedding plane about 2 meters square through a thickness of over 8 meters was sawn from the bedrock in columns 0.5 meter square (Figure 6). In addition, a borehole was drilled through approximately 180 meters of the sequence and continuous core was obtained.

The analysis of the sawn columns showed that the rock is not uniform throughout. When the surfaces cutting through the beds were cleaned, the nature of the layers was clearly visible, revealing variation in the grain size of the sediment (Figure 7). The thickness of successive layers also varies significantly. Analysis revealed patterns of sedimentation, including rapid depositional events separated by periods of

Figure 6. Excavation of slate from the Eschenbach-Bocksberg quarry, Bundenbach, as part of the Nahecaris Project.

Figure 7. Sawn surfaces of slate from where columns were extracted in the Eschenbach-Bocksberg quarry, Bundenbach.

slow background accumulation. Tectonic activity has distorted the beds in places, and introduced veins of quartz.

The operation was difficult, not least because the layers in the quarry are oriented nearly vertically. The extracted columns were carefully split and the sedimentology and fossil content were documented (Figure 8). Some parts of the sequence were virtually barren of fossils, whereas they were relatively common elsewhere, particularly in two newly discovered layers. More than ten thousand fossils were recorded, although only about one in a thousand is well preserved.

The 1997 excavation provided material for several extensive scientific studies, resulting in more than thirty publications. The investigation of the block of slate has not yet been completed.

Figure 8. A stack of split slate from the column extracted from the Eschenbach-Bocksberg quarry, Bundenbach.

THE ANIMALS OF
THE HUNSRÜCK SEA

FIXED BOTTOM DWELLERS (SESSILE BENTHOS)

Sponges

Corals

Brachiopods and Clams

Cystoids and Mitrates

Blastoids

Edrioasteroids

Crinoids

 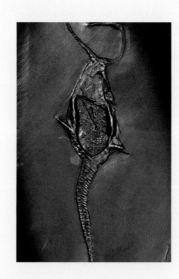 

Sponges (Porifera)

Sponges are primitive animals that have no organ systems. Nevertheless, they respond to environmental stimuli by slowing or stopping the flow of water through the body. In essence, sponges consist of cells with different functions that work in concert and are often supported by a skeleton that may be mineralized in silica or calcium carbonate. Sponges have colonized many habitats and are often found in very large populations. They are present in all seas, and they sometimes also occur in freshwater. They are common in reef settings, where they show a variety of growth habits and a range of spectacular colors. They can reach very high densities in the deep sea, because there they have fewer competitors for food than they do in shallower water. There are few sponges in the Hunsrück Slate, and even fewer are preserved complete. Several different forms are known, all of them with siliceous skeletons (Figures 9 to 12). Most of the sponges remain to be investigated in detail. Even after millions of years, although their colors are no more, many Hunsrück Slate sponges are still strikingly beautiful.

"Protospongia"

This sponge fossil resembles an artistically designed vase or a basket. It is the most common Hunsrück Slate sponge, although most specimens are only fragments. It shows some similarities to *Protospongia,* a sponge that occurs in the famous Burgess Shale and many other Cambrian deposits. The Hunsrück Slate specimens probably represent a new genus, but the material is in need of detailed investigation, and for the time being we refer to it as "Protospongia." The skeleton is composed of siliceous spicules, which are arranged in a rectilinear pattern. Pyritization of the spicules creates an elaborate golden meshwork.

Like their descendants living today, the Hunsrück Slate sponges fed on microorganisms, which they filtered from water passing through the pore openings. Small flagella concealed in numerous chambers generated a current, from which the smallest food particles were captured. Remarkably, brittle stars are often found inside the open vase: two brittle stars of the genus *Furcaster* lie inside the incomplete sponge on the right of one slab shown here (Figure 10). They presumably fed on materials carried by the flow of water through the sponge. They were also protected inside the sponge from potential dangers including predators.

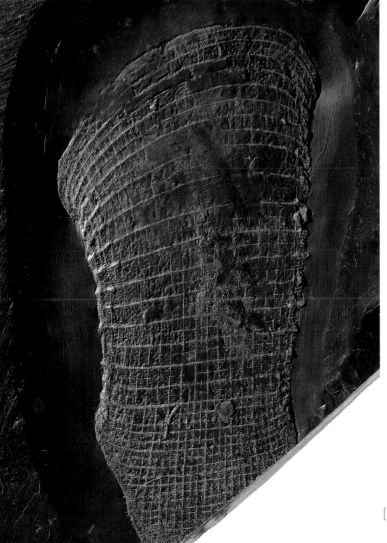

Figure 9. A nearly complete example of *"Protospongia" rhenana* (height about 20 centimeters).

Figure 10. Slate slab preserving a complete and partial example
of *"Protospongia" rhenana*, the latter containing two brittle stars.
The specimens show the network of spicules typical of this sponge
(height of the slab 26 centimeters).

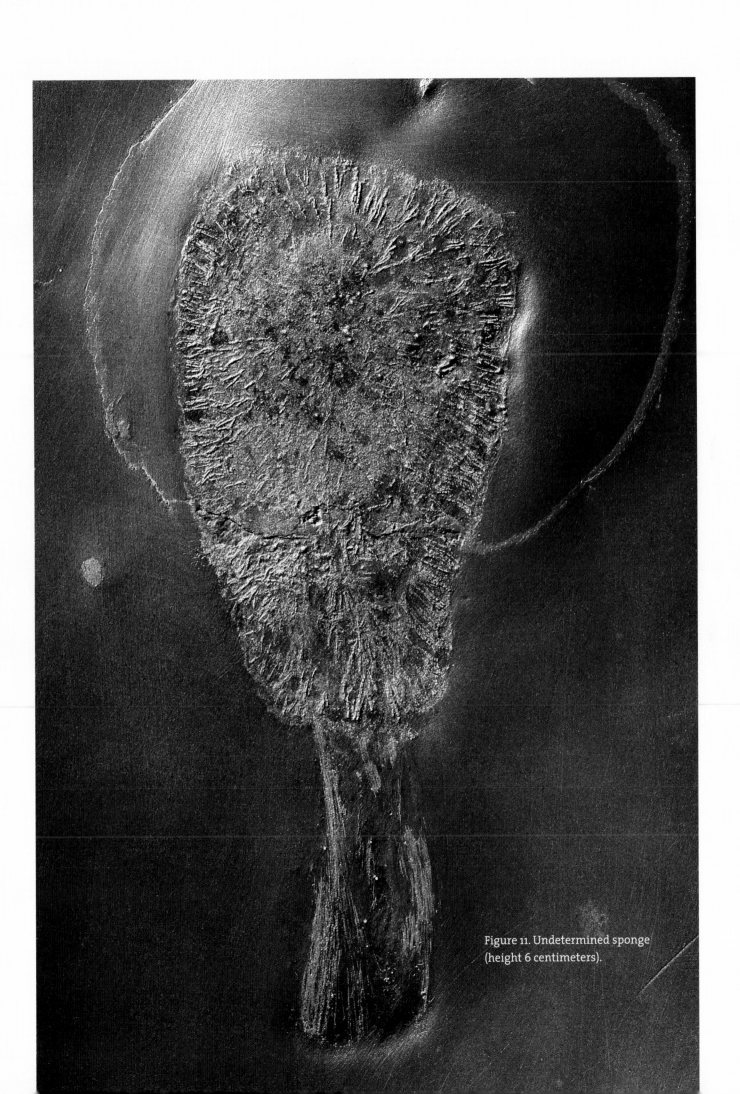

Figure 11. Undetermined sponge (height 6 centimeters).

Figure 12. Undetermined sponge
(height 7 centimeters).

Sponges on *Mimetaster*

Where the seafloor was muddy and soft, sponges had to adapt in order to anchor safely and avoid sinking in the sediment. A species of sponge from the Hunsrück Slate evolved a very special way of achieving this: it grew on the head shield of the bizarre-looking arthropod *Mimetaster hexagonalis* (Figure 13). It has long been noticed that isolated sponge spicules are commonly found associated with this small arthropod. A spectacular slate slab with a mass accumulation of *Mimetaster hexagonalis* clearly shows small sponges attached to the arthropods (see Figure 58). Once safely settled on the back of *Mimetaster*, the sponge always remained above the muddy seabed and was carried by the arthropod to different places. *Mimetaster* presumably used the many small limbs along its trunk to stir up the mud and release food particles, some of which were captured by the sponge. But the arthropod also benefited from this peculiar symbiosis, because it was concealed from potential predators by the sponge.

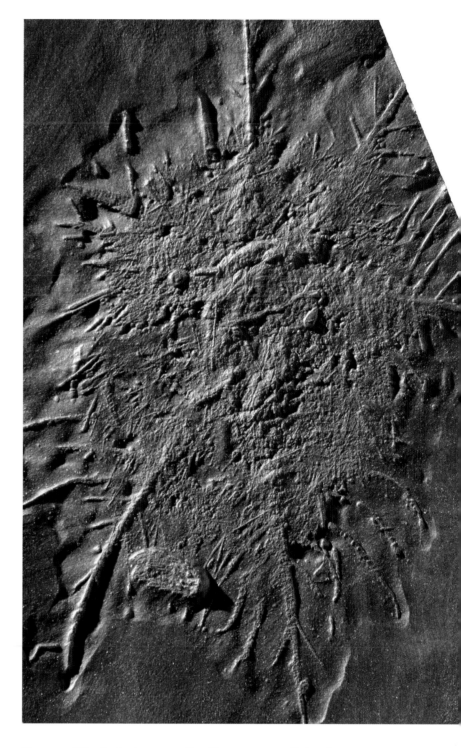

Figure 13. *Mimetaster hexagonalis* with a sponge on its head shield. The six rays of the arthropod project laterally from under the sponge body (diameter about 6 centimeters).

Corals (Anthozoa)

Corals are relatively common at most sites in the Hunsrück Slate, but only four forms (*Zaphrentis, Volgerophyllum, Pleurodictyum,* and *Aulopora*) have been mentioned in the literature so far. Other types have been found, including *Favosites,* which also occurs in the Devonian reefs of the Eifel region to the north and is common in older reefs ranging back to the Upper Ordovician (Figure 14).

The different species of Hunsrück Slate corals are distinguished by the fine structure of the skeleton, details that can be seen only in polished sections of the best-preserved fossils. Even though specimens often appear three-dimensional, dissolution and pyritization may have obscured the structure of the skeleton. The rugose coral *Zaphrentis* is the most common; with its slightly curved apex and circular cross-section this coral looks like a hat (Figure 15). Unlike reef-building corals, rugose corals like *Zaphrentis* usually occur as single individuals (Figure 16), sometimes clustered on shells, often those of cephalopods. The skeletons of these corals were so robust that, in contrast to many of the Hunsrück Slate animals, *Zaphrentis* was fossilized in great numbers and survived the subsequent tectonic history and deformation of the sedimentary rocks in the Hunsrück region.

The small colonies of the coral *Pleurodictyum* are less common. They often occur in more sandy layers (Figure 17) and are

Figure 15. Three-dimensionally preserved rugose coral *Zaphrentis* (height 3 centimeters).

frequently found where sandstone is interbedded with slate (formerly mud). In the central Hunsrück region sandy layers are abundant in the higher parts of the Hunsrück Slate sequence, and they are more prevalent within the roof slates in the middle Rhine and Wisper regions. The underside of *Pleurodictyum* often reveals the tube of a worm that lived inside the small colony (Figure 18). The top shows the honeycomb pattern typical of corallites of this genus.

Large numbers of individual rugose corals are sometimes found aligned, perhaps along some former attachment surface (Figure 19). Among the corals from the Hunsrück Slate that remain to be described and identified are relatively large colonies that encrust shells, often covering them completely, as in the case of the elongate shell of a cephalopod shown here (Figure 20). Such encrusting corals have occasionally been confused with sponges, even though sponges are very different in appearance.

Figure 14. *Favosites* sp. (width of colony about 11 centimeters).

Figure 16. Undetermined rugose coral (diameter 5 centimeters).

Figure 17. A small colony of the tabulate coral *Pleurodictyum* viewed from the underside (diameter 3.5 centimeters).

Figure 18. *Pleurodictyum problematicum* with worm tube, from the Devonian strata of the Eifel (diameter 3 centimeters).

Figure 19. Numerous individual corals encrusting an undetermined firm substrate, perhaps the remains of an alga or sponge (the preserved length is 40 centimeters). The opening of the corals points in the same direction.

Figure 20. Colonial coral encrusting a cephalopod shell (a straight nautiloid) lying on the seabed (length of cephalopod 20 centimeters).

Brachiopods (Brachiopoda) and Clams (Bivalvia)

The bivalved shells of brachiopods and clams are quite similar in appearance and can even be confused at first glance. But these two groups are distinguished by their different plane of symmetry. The plane of symmetry of brachiopods passes through both shells (the valves are dorsal and ventral), whereas in clams this plane is usually between the shells (which are left and right). This means that the shells of clams are mirror images of each other, but the shells of brachiopods are not. The soft anatomy of these animals also differs radically—indeed they belong to different animal phyla. As the name "brachiopod" (literally "arm foot") indicates, brachiopods (Phylum Brachiopoda) have two fleshy arms inside the shell that bear tentacle-like structures, which function in filter feeding and oxygen exchange. Brachiopods are usually attached to the substrate by a stalk known as the pedicle. Over 30,000 species of fossil brachiopods are known, but there are only about 380 species living today. Clams, on the other hand, are much more diverse today, with about 20,000 living species. They belong to the Phylum Mollusca, which also includes snails and cephalopods. Clams and brachiopods some-

Figure 22. Undetermined large clam (diameter 12 centimeters).

Figure 21. The clam *Panenka* sp. (diameter 6.6 centimeters).

times live in the same habitats and exploit similar food sources. Clams (Figures 21 and 22) were much less abundant and diverse than brachiopods during the Devonian, and brachiopods are more common in the Hunsrück Slate (Figures 23 and 24). They are often overlooked by collectors, however, because they are usually small and inconspicuous.

One of the most spectacular fossils discovered since slate mining became uneconomical and ceased at Bundenbach in 1999 features several animals, including brachiopods (Figure 25). In addition to a large, well-preserved specimen of the sea star *Echinasterella*, this slate slab contains three brachiopods preserving not only the shell, but also the long, wrinkled stalk, which had muscular soft tissue covered by a tough skin.[15] The x-radiograph clearly shows the largest of these brachiopods to the left of the brittle star. The two valves of the shell gape slightly, and the stalk shows an S-shaped curve near its extremity. The

Figure 23. Undetermined spiriferid brachiopod (width 7 centimeters) from the Hunsrück Slate of the Eifel.

Figure 24. Slab with brachiopods and large tentaculitids (cone-shaped shells), from the Hunsrück Slate at Geroldstein in the Wispertal area (width of the slate slab 18 centimeters).

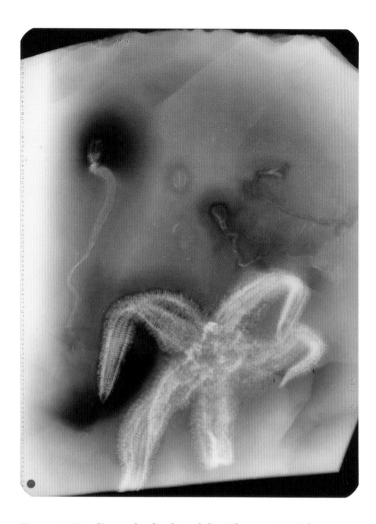

Figure 25. X-radiograph of a slate slab with a sea star *Echin-asterella*, three lingulid brachiopods with stalks, two juvenile specimens of the arthropod *Vachonisia rogeri*, crinoids, and plant debris (diameter of *Echinasterella* 12 centimeters).

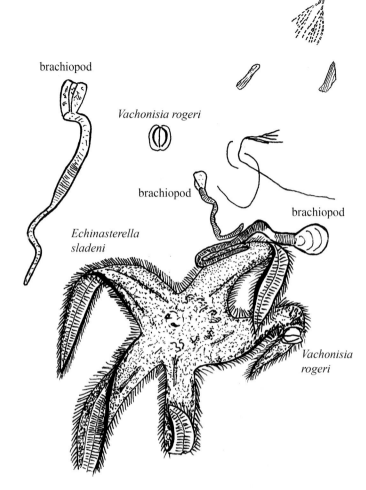

Figure 26. Schematic drawing of the slate slab shown in Figure 25.

other two brachiopods are above and to the right of the brittle star (Figure 26). The brachiopods on this slate slab are lingulids. They can be compared to living representatives, such as modern-day examples from Queensland, Australia (Figure 27). The slab also preserves two juvenile individuals of the arthropod *Vachonisia rogeri* (p. 62) and two fragments of crinoids, their position indicated in the schematic drawing. This fossiliferous slab provides an example of the remarkable preservation that occurs in parts of the Hunsrück Slate.

Figure 27. Recent brachiopods *Lingula anatina* from Geof-frey Bay, Magnetic Island, Australia (shell length about 4 centimeters).

Cystoids (Cystoidea) and Mitrates (Stylophora)

Cystoids and mitrates are echinoderms, relatives of crinoids, sea stars, and sea urchins. Cystoids and mitrates are known only in marine deposits of Paleozoic age. They had a plated skeleton with appendages projecting from the body.

Two forms from the Hunsrück Slate, the cystoid *Regulaecystis pleurocystoides* and the mitrate *Rhenocystis latipedunculata*, are not closely related but evolved a similar flattened body and long flexible stem or tail (Figures 28 and 29). They used this tail to move on the sediment surface or perhaps to cling to an appropriate object (see Figures 115 and 116). *Regulaecystis* fed with the paired arms; the mode of feeding in *Rhenocystis* is uncertain. For a time some paleontologists regarded mitrates such as *Rhenocystis* as closely related to the earliest vertebrates. We now know that this is not the case even though echinoderms and vertebrates belong together in a larger group known as deuterostomes.

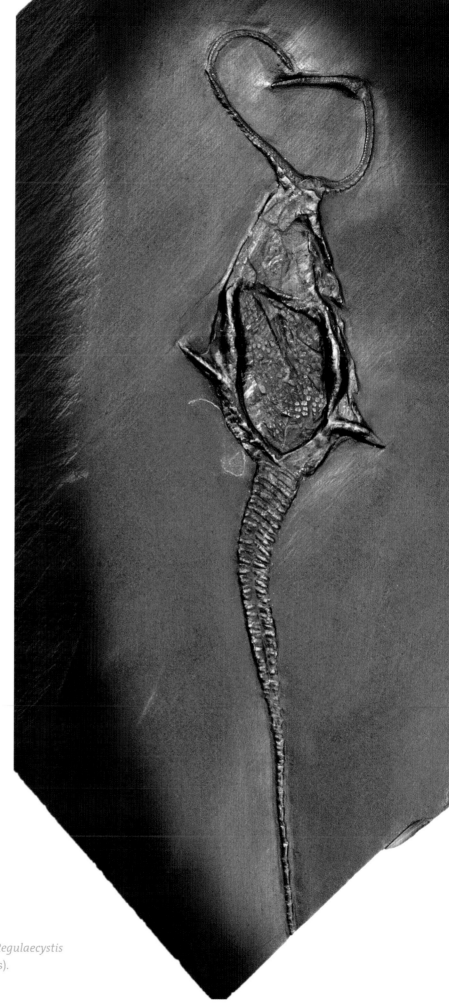

Figure 28. Nearly complete specimen of a *Regulaecystis pleurocystoides* (total height 13 centimeters).

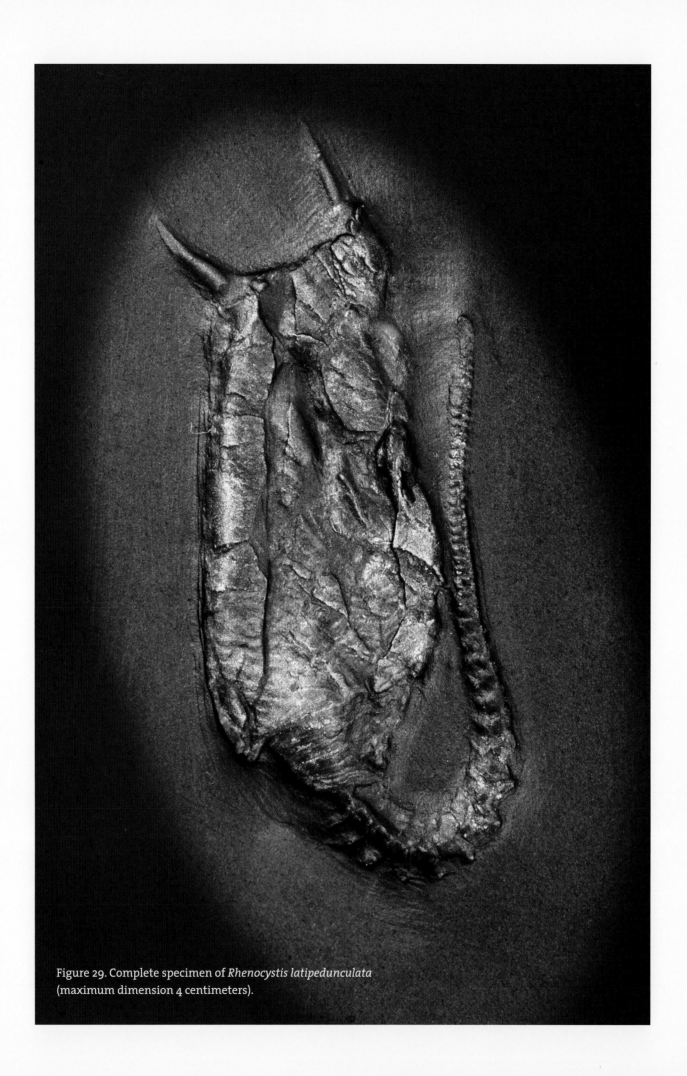

Figure 29. Complete specimen of *Rhenocystis latipedunculata* (maximum dimension 4 centimeters).

Blastoids (Blastoidea)

Blastoids are echinoderms with some similarities to the sea lilies or stemmed crinoids. Although they were widespread during the Paleozoic, they are one of the rarest groups in the Hunsrück Slate. *Schizotremites osoleae,* a blastoid, was described by Walther Lehmann in 1949.[16] An x-radiograph shows the ovoid plated body and the slender stem that attached it to the muddy seafloor (Figure 30). In the photograph shown here, these features are concealed by the multiple arms (called brachioles), which give the specimen the appearance of a rather unruly hairstyle (Figure 31). *Schizotremites osoleae* filtered small food particles from the sea and transported them down grooves in the arms to the mouth. Blastoids with the arms in place are very rare, emphasizing the important role that rapid burial plays in the exceptional preservation of the Hunsrück Slate fossils.

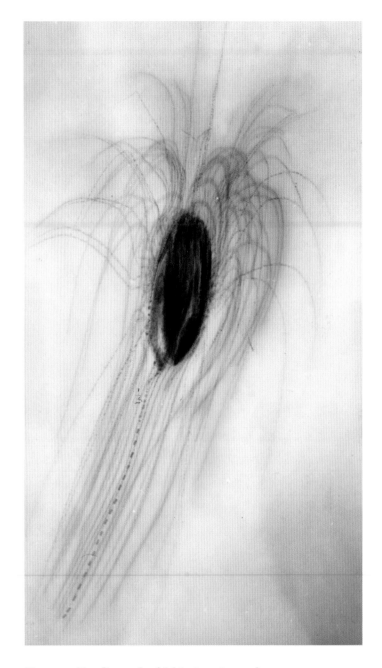

Figure 30. X-radiograph of *Schizotremites osoleae* (height 7 centimeters).

Figure 31. *Schizotremites osoleae*
(height 7 centimeters).

Edrioasteroids (Edrioasteroidea)

A remarkably detailed x-radiograph of a tiny specimen of *Pyrgocystis coronaeformis* led to some confusion in 1990.[17] At first it was mistakenly interpreted as a new wormlike organism living in its own secreted tube. In part this was due to a general problem in interpreting x-radiographs, the difficulty of reconstructing a three-dimensional animal based on an image in two dimensions. Further inspection of specimen and x-radiograph showed it to be only the second example of *Pyrgocystis* discovered, although better preserved examples have been found since (Figures 32 to 34). *Pyrgocystis coronaeformis* is the only representative of the edrioasteroids, an unusual extinct group of echinoderms, in the Hunsrück Slate. It is also one of the rarest animals in the fauna—fewer than twenty specimens are known. The top or oral side shows a fivefold symmetry radiating from the central mouth. The stalk is a flexible tube covered with scalelike plates. A saclike structure with many tiny scattered plates anchored the animal in the sediment.

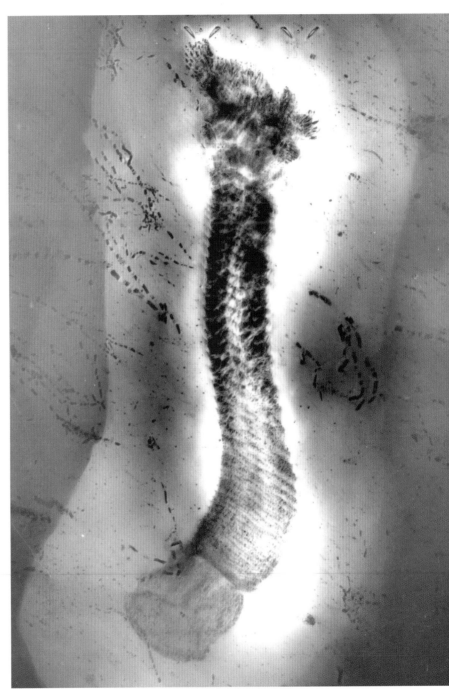

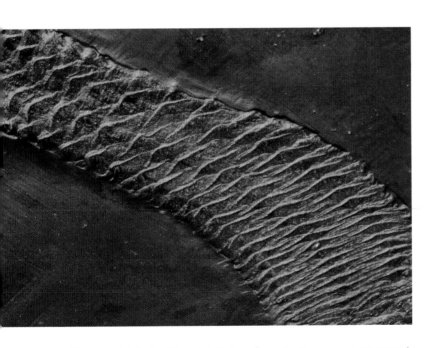

Figure 32. Detail of the plated stem (length shown, 5 centimeters) of *Pyrgocystis coronaeformis.*

Figure 33. X-radiograph of *Pyrgocystis coronaeformis* (height 10 centimeters).

Figure 34. Complete specimen of *Pyrgocystis coronaeformis* (height 10 centimeters).

Crinoids (Crinoidea)

Crinoids, commonly known as sea lilies, are elegant creatures, which evolved more than 500 million years ago. The common name refers to their familiar form, a stem with a crown, the crown consisting of a cup-shaped structure with a number of arms (Figure 35). The arms, which carry tentacle-like features called tube feet that filter suspended particles from the water, give the animals their flowerlike appearance. Many stalked crinoids are permanently anchored, although some can change their position. Some living stalked crinoids, for example, can crawl along the ocean floor using their arms. As well as the stalked sea lilies, there are free-living crinoids called feather stars, which have shorter stems during the early stages of growth and can use their arms to swim. The free-living crinoids account for some 550 of about 620 species of living crinoids; today's stalked crinoids are essentially confined to the deep sea.

With about 70 described species, crinoids are exceptionally diverse in the Hunsrück Slate. All these forms are stalked; stemless crinoids did not evolve until the Mesozoic. Many crinoids normally attach to hard surfaces. Where the seafloor is covered with sediment, they are often attached to shell fragments or to other animals, such as sponges, that live on the substrate. Several forms have adapted to live on muddy seafloors by evolving a much-branched root tuft that attaches the stem to the sediment. In other species up to two-thirds of the stem, which grows lateral branches like those of a tree, remains in contact with the substrate, while the rest, including the crown, is elevated above the seabed. All growth stages from tiny juveniles to adult animals are represented in the Hunsrück Slate. Both individual specimens and groups (Figure 36) are preserved, some of which are spectacular in appearance. The preservation of such a diversity and abundance of complete crinoids is particularly rare because they normally decay very quickly after death, disarticulating into thousands of skeletal elements that are easily scattered by bottom currents. The remarkable assemblage of crinoids in the Hunsrück Slate is the result of rapid burial by turbidity currents. The crinoids alone could fill an entire book, so we are exhibiting only some of the finest specimens here.

Figure 35. A crown about 10 centimeters high of the very rare crinoid *Propoteriocrinus scopae.*

Figure 36. A cluster of specimens of *Thallocrinus* sp. A small brittle
star is stranded on the crown of the specimen at the top of the slab.
The average diameter of the crowns is about 4 centimeters.

Figure 37. A specimen of *Gastrocrinus giganteus* showing a particularly well preserved anal sac in the middle of the cup and cirri projecting from the stem (stem and crown together about 25 centimeters high).

Gastrocrinus

Gastrocrinus giganteus is a relatively rare species in the Hunsrück Slate. The stem of this crinoid is particularly long and robust (Figure 37). Upwardly directed tapering appendages called cirri project at regular intervals from the stem. The crown bears a massive plated structure called the anal sac. The anus was at the summit of this structure, which helped to separate the waste products from the filter feeding arms. The spines at the top of the anal sac may have prevented other invertebrates from attaching there, either as parasites or as suspension feeders. The detailed image shows that the ends of the arms of *Gastrocrinus* branched almost like a tree, giving the specimen a plantlike appearance (Figure 38).

Figure 38. Detail of the crown of *Gastrocrinus giganteus* showing the branching of the arms (width of the area illustrated 10 centimeters).

Hapalocrinus

Several species of *Hapalocrinus* are present in the Hunsrück Slate. It is one of the more common crinoids, but that makes it no less striking. Some species of *Hapalocrinus* have very long arms, which bear up to thirty small branches (pinnules), so that the crown appears to be very large. Different growth stages are represented in the group shown here, with younger, smaller individuals attached to the stems of older ones (Figure 39). *Hapalocrinus* is found in both small and large groups. The individuals on this slate slab were buried at different levels in the sediment, giving the group a very lifelike appearance. You could almost forget that these animals are some 400 million years old! In this case, a slow-moving cloud of suspended sediment overwhelmed the crinoids, but even after compaction of the mud to slate they occupy a thickness of up to 20 millimeters. The stems appear to be converging on a single attachment spot off the edge of the slab. The group was presumably pushed over and buried alive more or less where they were living on the seafloor.

Figure 39. Slate slab with about 20 individuals of *Hapalocrinus innoxius*, some *Triacrinus koenigswaldi,* and two brittle stars *Eospondylus primigenius* (slab about 50 centimeters high).

One particularly fine example from the Hunsrück Slate is a group of eight specimens of *Hapalocrinus frechi*. The arms divide close to the cup, forming ten strong attachments, and then divide again. These branches bear a series of straight spines and many pinnules, forming a fine net to trap tiny food particles (Figure 40). The food particles were transported to a trough at the base of each arm with the help of small tentacles, or tube feet, and carried to the mouth at the center of the cup. The strong spines at the bottom of the cup may have functioned to ward off parasites. The largest specimen of the group has a stem about 20 centimeters long and arms about 7 centimeters long; the arms in the smallest specimen measure only 1.5 centimeters (Figure 41).

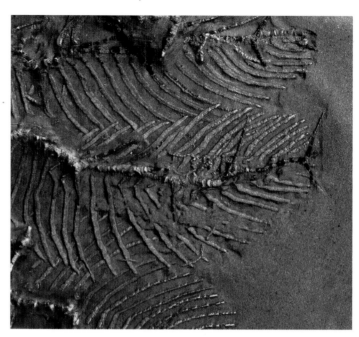

Figure 40. Detail of *Hapalocrinus frechi* showing the end of a spiny arm branch, and many curved pinnules (height of the area illustrated 2 centimeters).

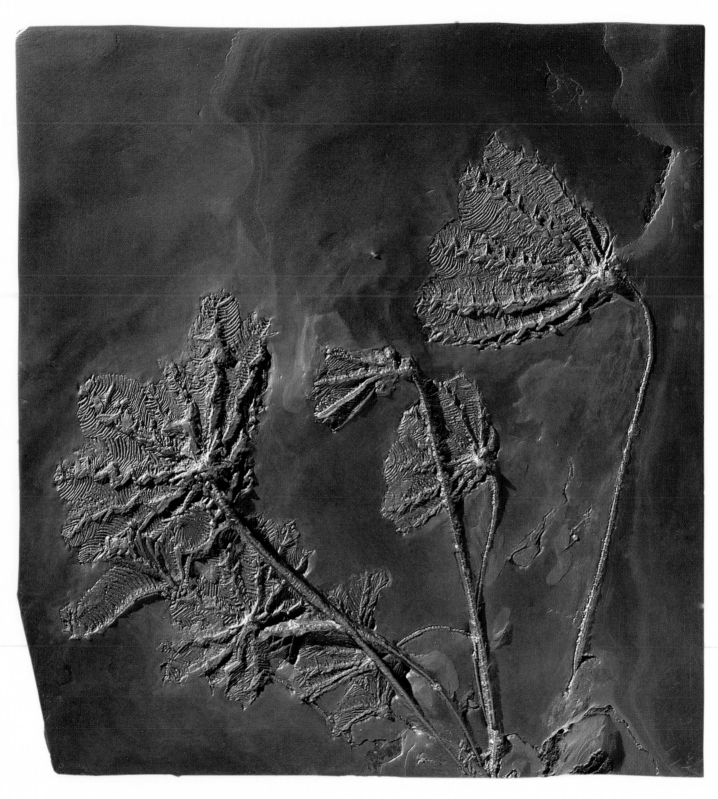

Figure 41. Slate slab with eight individuals of *Hapalocrinus frechi* (height of the crown on the right side 6 centimeters).

Dicirrocrinus with a Juvenile of *Codiacrinus*

The highly magnified image of one crinoid, *Dicirrocrinus comtus,* shows that it was not alone! A juvenile of *Codiacrinus schultzei* has attached itself to a branch of the stem of the *Dicirrocrinus* (Figure 42). The small individual shown here is only about 5 percent of the size to which *C. schultzei* grew as an adult. The stem and arms show just a few of the rounded barrel-like plates that make up the mature skeleton. Both juvenile and adult crinoids lived in the same environment, and different generations are often found together on the slate slabs that represent the Hunsrück seafloor.

Viewing this juvenile in its larger context, it is evident that the slate slab preserves only a portion of the stem of *Dicirrocrinus comtus* (Figure 43). It is the lower part of the stem, which was often partially buried in the sediment to provide support and anchorage when the animal was alive; the crown has been lost.

Acanthocrinus

One of the most spectacular fossils found in the Hunsrück Slate was discovered and described toward the end of the nineteenth century. Otto Jaekel named this new species *Acanthocrinus rex* (the king).[18] Specimens like this, 40 centimeters in dimension, completely undamaged and suitable for preparation, were very rarely found (Figure 44). It came from a slate mine near Kaub on the Rhine. Kaub gives its name to the slates of the middle Rhine and the Hunsrück region, where the rock sequence is referred to as the Kaub Formation.

The end of the stem of Jaekel's specimen is spirally enrolled and attached to a shell. The stem consists of thick robust plates separated by thinner ones. The lower plates of the cup bear strong spines, which are positioned to discourage parasites. The five arms of the crown each divide three times close to their attachment to the cup, resulting in a total of forty branches that, with their long pinnules, look like palm fronds. The crown of the crinoid forms a netlike structure, which filtered small food particles suspended in the seawater. To the right of the stem is a small blastoid, *Pentremitidea medusa,* a stalked form related to crinoids. Three juvenile blastoids are also present on this slab, but are not evident at this magnification. For a long time these specimens were the only blastoids known from the Hunsrück Slate.

Unfortunately, this magnificent piece was lost during the Second World War, and the image here is reproduced from Jaekel's original.

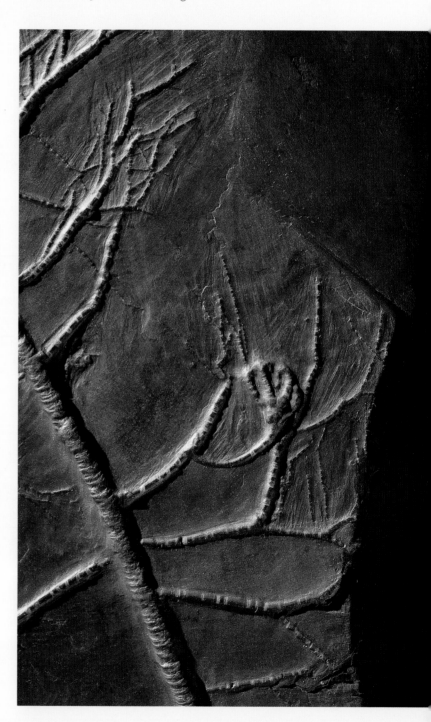

Figure 42. Juvenile stage of the crinoid *Codiacrinus schultzei* attached to the root of *Dicirrocrinus comtus.* The specimen of *C. schultzei* is 3 centimeters long.

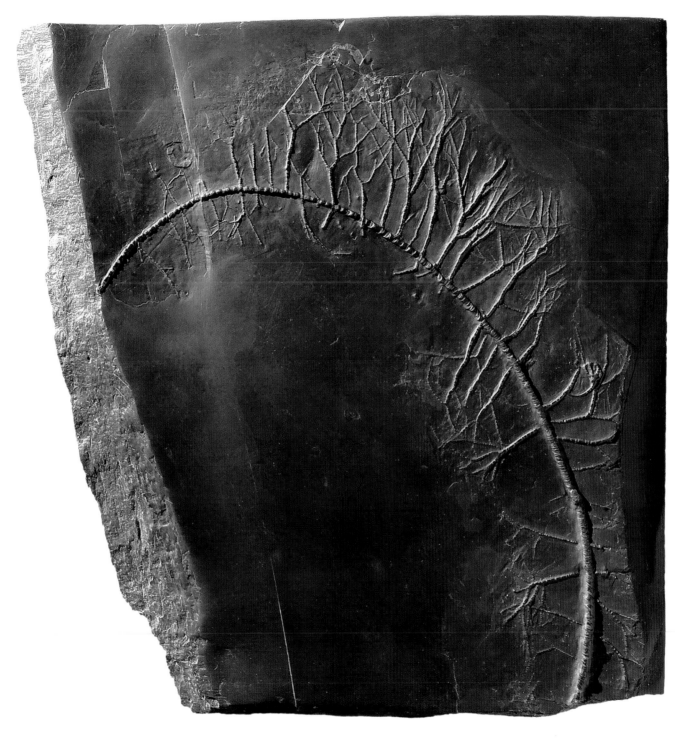

Figure 43. A 15-centimeter length of the stem of the crinoid
Dicirrocrinus comtus, showing the branching "roots."

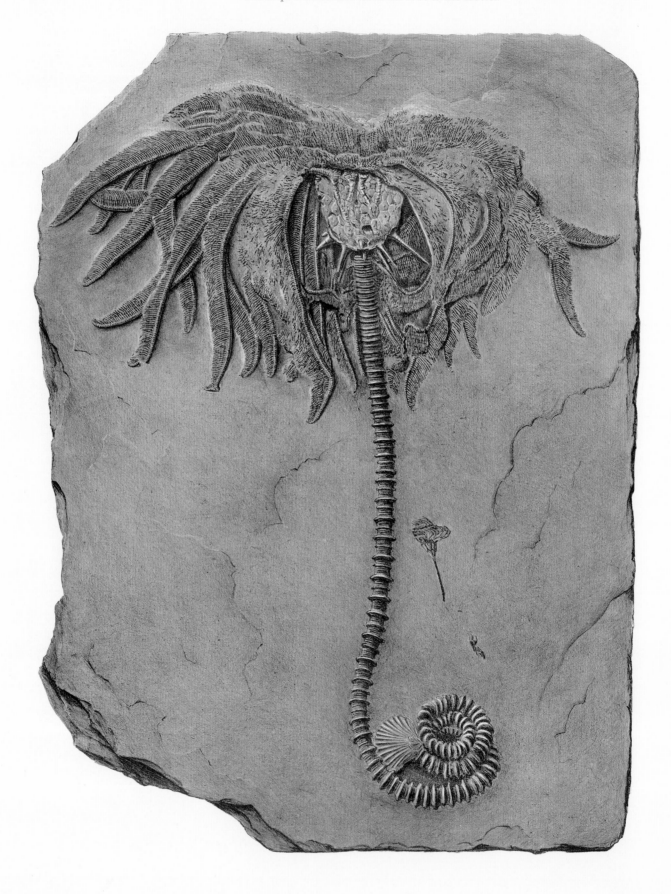

Figure 44. Lithograph of *Acanthocrinus rex*, from Jaekel (1895). The crinoid measures about 40 centimeters from the root to the crown.

Imitatocrinus and *Bathericrinus*

The crinoid *Imitatocrinus gracilior* has a crown that includes a large cup consisting of two circles of plates (Figure 45). The cup and the top of the stem bear spines to combat parasites. Close to the cup, the arms divide into two branches that zigzag near their ends, where they too bear strong spines. A view of the entire slab reveals that both this specimen of *Imitatocrinus gracilior* and one of *Bathericrinus semipinnulatus* are attached to the shell of a straight nautiloid (a cephalopod, like modern *Nautilus* and squids), which provided the hard surface they required for attachment (Figure 46). *Bathericrinus,* the individual to the right, is one of the rarer crinoids in the Hunsrück Slate. The branches from the lower part of the stem served as roots, and prevented the animal from sinking into the soft mud. *Bathericrinus* has a small cone-shaped cup, and the arms divide into two branches, each bearing long lateral branches alternately on one side and the other. Other decaying animals and shell fragments carried by bottom currents have accumulated against the cephalopod shell and become snagged on the stems of the crinoids.

An extension crack caused by tectonic processes traverses the slate slab. It became filled with silica, which precipitated from fluids, and was later transformed into milky quartz. The slate separated on either side of the fracture, displacing parts of the crown of *Imitatocrinus gracilior* relative to each other. So in addition to evidence of life in the Hunsrück sea, this slate slab reveals how mountain-building processes have affected the rock sequence.

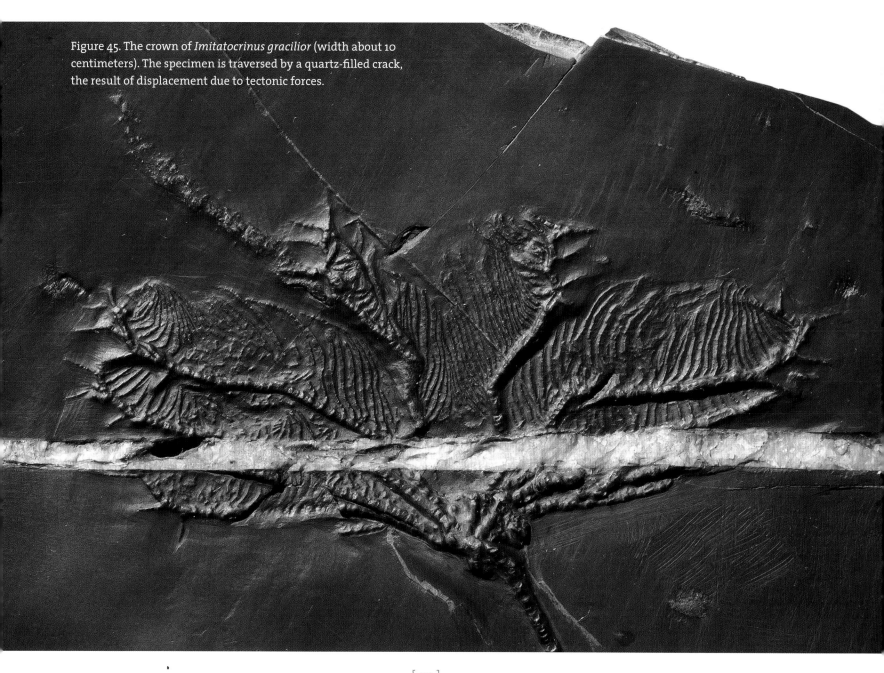

Figure 45. The crown of *Imitatocrinus gracilior* (width about 10 centimeters). The specimen is traversed by a quartz-filled crack, the result of displacement due to tectonic forces.

Figure 46. The crinoids *Imitatocrinus gracilior* and *Bathericrinus semipinnulatus* attached to the remains of a nautiloid shell (width of the fossil group 30 centimeters).

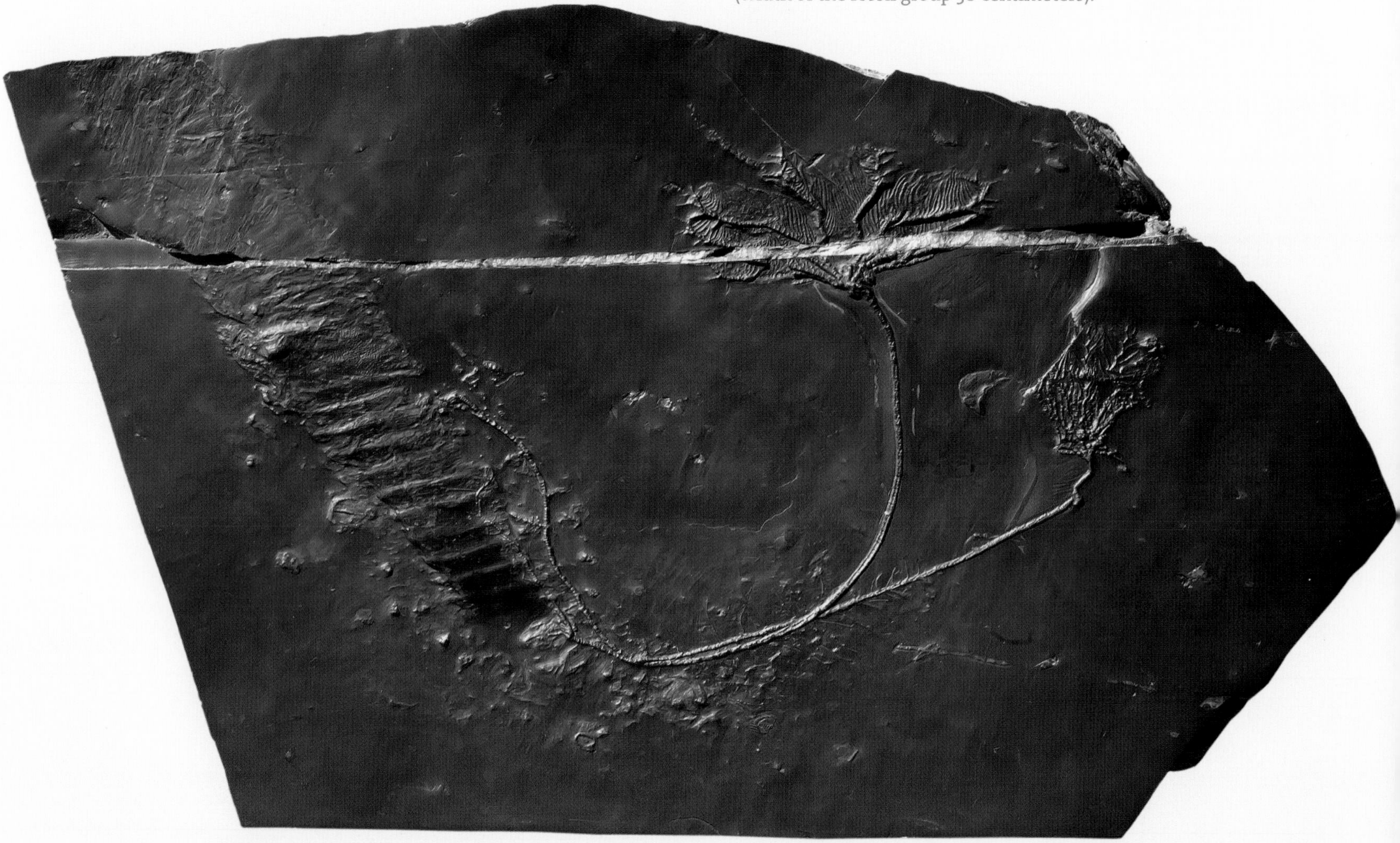

Taxocrinus

Taxocrinus is one of the largest crinoids found in the Hunsrück Slate, with stems reaching well over 1 meter in length. The specimen shown here measures about 50 centimeters, a long way from its potential size (Figure 47). The crinoid drifted through the water as a larva, and settled on a solitary rugose coral, which may have been alive at the time. As the crinoid grew, the lower half of the stem remained lying on the seabed, where it branched into a system of roots. These extensions of the stem anchored the crinoid and prevented it from sinking into the mud. The small cup consisted of three circles of small plates as well as the lowest plates of the arms. The arms divided four times, and often curved inward at their end. Another small crinoid (*Hapalocrinus*) is attached about halfway along the stalk of the *Taxocrinus,* apparently where the portion of the stalk with "roots" emerged from a thin layer of sediment. The specimen was prepared from below in terms of its position on the Hunsrück seafloor, and the sediment layers can be identified on the basis of slight differences in color. Only the last third of the stem was exposed above the sediment surface when the crinoid was killed by an influx of mud, as evidenced by the bend at this point. The crinoid could not escape from the rapidly deposited sediment and presumably suffocated. The solitary coral in the upper right of the photo was also buried by this event. Specimens of *Taxocrinus stuertzi* of this size were very rarely found complete. Commercial slate was almost always sawn into blocks before splitting, and these blocks were normally smaller than 35 by 45 centimeters.

Figure 47. The crinoid *Taxocrinus stuertzi* was buried in three successive layers of mud. Even though these layers have now turned to slate, the different horizons can be distinguished by color. The total length of the crinoid is 50 centimeters.

MOBILE BOTTOM DWELLERS (VAGRANT BENTHOS)

Tentaculitids

Bristle Worms

Arthropods

The Sea Urchin *Rhenechinus*

Brittle Stars and Sea Stars

Tentaculitids (Tentaculitida)

Tentaculitids are small and inconspicuous, but millions of specimens are preserved in the Hunsrück Slate and in many other Paleozoic deposits (Figures 48 and 49). These small animals, with their cone-shaped shells that are usually only a few millimeters long, are still a puzzle to paleontologists—it is not clear to what group they belong. They are so numerous and diverse in form that they have been used for a long time to correlate sequences of sedimentary rocks, including those of Devonian age in Europe.

The soft tissues of tentaculitids have not been preserved, and their mode of life is not well known. The Hunsrück Slate, however, has provided a number of insights into the paleoecology of these problematic animals, which are being studied at the Steinmann Institute, University of Bonn. Specimens of the arthropod *Mimetaster hexagonalis,* for example, have been found with tentaculitids adhering to the spines of their extraordinary head shield (Figure 50). Presumably this relationship was commensal: the tentaculitids positioned themselves to take advantage of food particles that *Mimetaster* stirred up from the mud as it used its appendages to feed. The tentaculitids in the Hunsrück Slate of the Wispertal area in the Taunus Mountains are larger than most, reaching lengths of up to 8 centimeters, and they probably lived in areas where the substrate was relatively sandy (Figure 51). There they are often found associated with brachiopods and gastropods, and with the disarticulated bony plates of placoderm fishes. Tentaculitids occasionally settled on unknown hard structures or on various shells (Figures 52 and 53). Most, however, probably lived in or on the sediment, where they were able to change their orientation to feed. Many thousands are found on some bedding planes.

Figure 48. Detailed view of some tentaculitids from the Wispertal area (shell lengths 40 millimeters and 20 millimeters).

Figure 49. Detailed view of tentaculitids aligned along the growth lines of a clam shell (width of area illustrated 2.5 centimeters).

Figure 50. Tentaculitids associated with *Mimetaster hexagonalis* (width of arthropod 8 centimeters).

Figure 51. Large tentaculitids on a fossiliferous slab from the Devonian strata of Wispertal (area illustrated 9 by 6 centimeters).

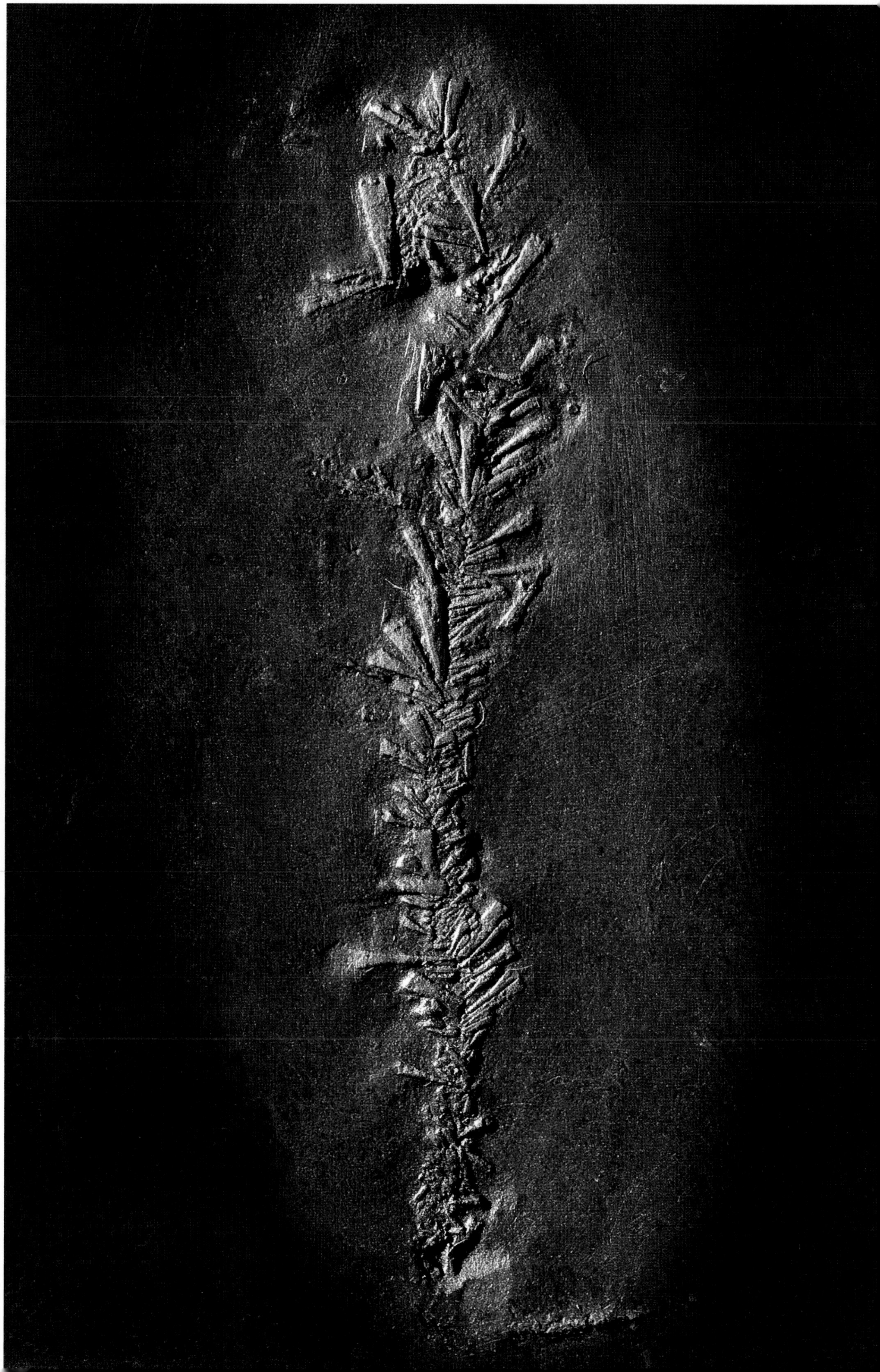

Figure 52. A group of tentaculitids aligned along a linear attachment (length 6 centimeters).

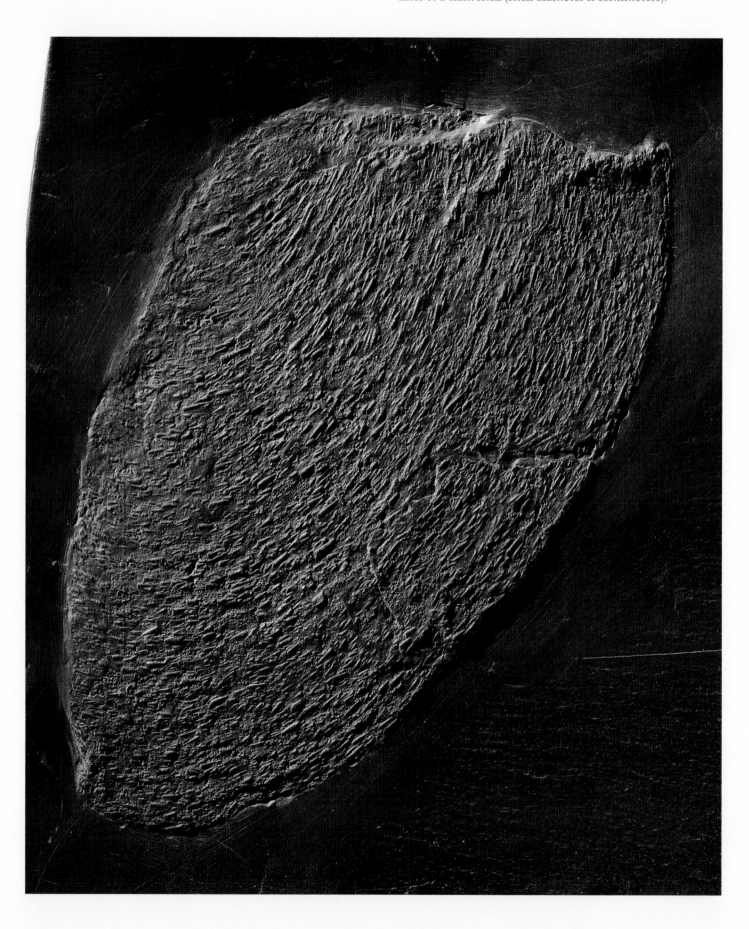

Figure 53. Tentaculitids aligned along the concentric growth lines of a clam shell (shell diameter 11 centimeters).

Bristle Worms
(Annelida, Polychaeta)

Figure 55. The bristle worm *Ewaldips feyi* (length 2 centimeters).

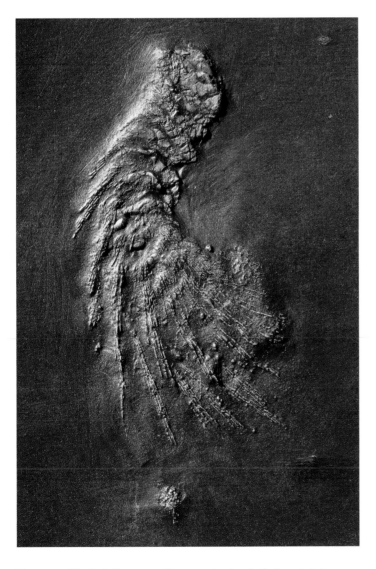

Figure 54. The bristle worm *Hunsrueckochaeta hohensteini* (length 4 centimeters).

Worms are often thought of as the obvious example of a soft-bodied animal, but polychaete worms, which are exclusively marine, have tough jaws and bristles on their limbs, both of which are somewhat decay resistant. The rest of the tissues, however, usually decompose rapidly, leaving no trace of the body outline. It requires very unusual conditions, such as those in the Hunsrück sea, to fossilize worms. The first reports of bristle worms from the Hunsrück Slate met with some skepticism because their preservation was so improbable, and of course other Hunsrück Slate fossils had been misidentified in the past. *Bundenbachochaeta*, published in 1995, clearly shows the structures of a bristle worm, and confirmed the presence of polychaetes in this famous deposit.[19]

With further research and preparation about twenty other specimens of bristle worms have been identified, and five distinct forms have been described and named. These include a more compact species with numerous and sometimes very long bristles on the appendages (Figure 54) and a longer form with many segments (Figure 55).

Other evidence of the activities of wormlike organisms both in and on the sediment includes strings of small fecal pellets made by unknown producers (see the discussion of *Palaeostella* on p. 91), and worm tubes that encrust shells or are made up of tentaculitids. Likewise, long, narrow wormlike strings of pyrite are often found in the Hunsrück Slate, but they do not preserve any details to indicate whether they even represent an animal, much less a particular kind of worm. Overall, we can assume that only a tiny fraction of the different kinds of worms that were present in the Hunsrück sea are preserved as fossils (Figure 56).

Figure 56. The bristle worm *Bundenbachochaeta eschenbachiensis,* with an x-radiograph of the fossil, and a reconstruction of the worm (length of the worm 4.8 centimeters).

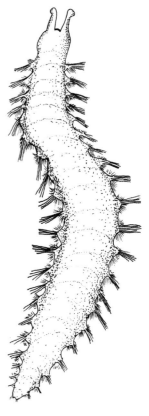

Arthropods (Arthropoda)

One of the most spectacular groups of fossils from the Hunsrück Slate is the arthropods, the group that includes today's crabs, insects, spiders, and millipedes. About 80 percent of all known living animals are arthropods. They are represented in all habitats. Even if most fossil arthropods are small and inconspicuous, they are important in understanding the evolution of the forms that populate our planet today. The arthropods living in the Hunsrück sea were very diverse. Some of them are recognizably similar to their modern relatives, while others are very different from any arthropod living today. This difference is especially marked in some specimens that were discovered in the Hunsrück Slate only recently, and represent groups that were previously thought to have gone extinct during the Cambrian, more than 100 million years earlier. Evolution seems to have been experimenting with body plans during the Cambrian explosion, when marine animals with shells and other kinds of mineralized skeletons first appeared. Thus we were surprised to find arthropods in the Hunsrück Slate that are similar to strange-looking forms that would be unremarkable in Cambrian deposits. Clearly the apparent loss of these arthropods in younger rocks resulted not from extinction, but from a lack of conditions suitable for their preservation. New, exceptionally preserved arthropods from the Silurian of England and the Ordovician of Morocco provide links between the unusual arthropods of the Hunsrück Slate and those of the Cambrian. The remarkable Hunsrück Slate arthropods *Mimetaster hexagonalis* and *Schinderhannes bartelsi,* for example, are clearly related to iconic Cambrian forms.

Marrellomorphs (Marrellomorpha)

The Marrellomorpha is a group of arthropods found in Paleozoic rocks. Only five genera are presently assigned to the group. *Marrella splendens* is known from thousands of specimens from the famous Middle Cambrian Burgess Shale in the Canadian Rockies. *Mimetaster hexagonalis,* an arthropod from the Hunsrück Slate that we met earlier in this book, is a close relative (Figure 57). *Mimetaster* is relatively common: Gabriele Kühl and Jes Rust provided a new description and interpretation in 2010 on the basis of more than 120 specimens.[20] However, it should be noted that this material was collected over a period of 160 years, and we have one single slate slab that preserves more than twenty individuals, the result of a mass burial (Figure 58). Closer inspection of this slab reveals that individual specimens are lying in different orientations to the bedding surface. Some are dorsal side up, others show the underside with many pairs of legs, and still others present various lateral views; all have been flattened in the slate (Figure 59). It is clear that the arthropods were transported in a turbulent cloud of suspended sediment and buried in different attitudes on the Hunsrück seabed. These different orientations were maintained as the arthropods were flattened and compacted in the slate.

The six-rayed head shield of *Mimetaster* looks like a star when viewed from above, so it is not surprising that collectors often refer to specimens as "Scheinstern," or false stars. They are often found in groups when the slate is split, and it is likely that they lived together on the seabed.

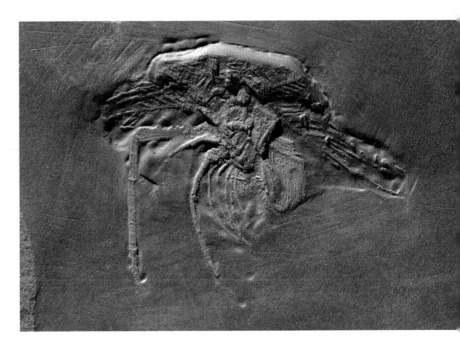

Figure 57. Frontal view of a *Mimetaster hexagonalis,* perhaps showing the attitude of the limbs in life (diameter 4 centimeters).

Figure 58. Slate slab with 22 mostly complete *Mimetaster hexagona-lis* specimens on a bedding plane, associated with spherical sponges (height of slab 25 centimeters).

Figure 59. Ventral side of a prepared *Mimetaster hexagonalis* preserving details of the limbs (diameter 5 centimeters).

Mimetaster is known only from the Hunsrück Slate, and this is also true of a very different representative of the marrellomorphs, *Vachonisia rogeri* (Figures 60 and 61). Kühl and her colleagues redescribed *Vachonisia* in 2008 based on twenty specimens, about 80 percent of those known.[21] The limbs are quite similar to those of *Mimetaster hexagonalis,* but the carapace is very different. Although *Mimetaster* looks superficially like a sea star, the carapace of *Vachonisia* viewed from above resembles the head shield of a crustacean, with a median ridge separating the two halves that cover the body. Indeed, many years ago, a specimen in the Bonn University collection was identified as a branchiopod crustacean, the group including living *Triops* as well as the so-called clam shrimps and water fleas. The specimen was long considered lost until one of us discovered it by chance where it had been misplaced in a museum drawer full of brachiopods, presumably due to a careless reading of the label! Several growth stages of *Vachonisia rogeri* are known, differing from one another in the shape of the carapace and the number of trunk limbs (Figure 62).

Figure 60. The marrellomorph *Vachonisia rogeri* exposed from the ventral side (length 7 centimeters).

Figure 61. A specimen of an adult *Vachonisia rogeri* prepared from the dorsal side (length 6.5 centimeters).

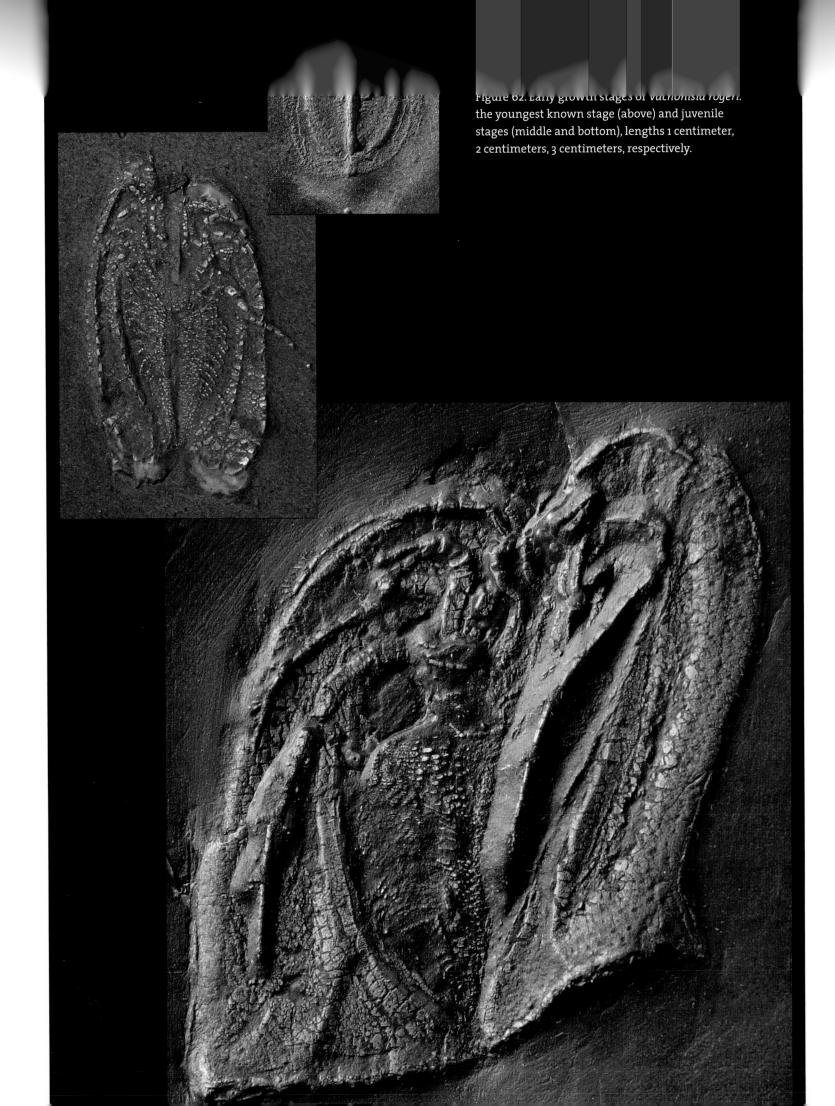

Figure 62. Early growth stages of *Vachonisia rogeri*: the youngest known stage (above) and juvenile stages (middle and bottom), lengths 1 centimeter, 2 centimeters, 3 centimeters, respectively.

Trilobites (Trilobita)

Trilobites range through the Paleozoic. The name "trilobite" comes from the longitudinal division of the exoskeleton into three lobes, but the body also consists of three sections: cephalon (head), thorax (trunk), and pygidium (tail). The dorsal exoskeleton is biomineralized, which gives it a high fossilization potential (Figure 63). The eyes of trilobites were compound, like those of many arthropods including insects and horseshoe crabs. But the lenses in trilobites were calcified, each lens consisting of a single calcite crystal, in contrast to those of living arthropods, which are composed of organic material. In the 1970s the collector, preparator, and paleontologist Günther Brassel performed the remarkable feat of revealing the individual lenses of the eye of the Hunsrück Slate trilobite *Chotecops*, with their central core, using only needles to remove the slate concealing them. The eyes of other species are made up of more than 15,000 individual lenses, and it is clear that trilobites had excellent vision. Such well-developed eyes indicate that light reached the Hunsrück seabed. Based on a comparison with the oceans today, it is unlikely that the depth of the water exceeded 200 meters.

Trilobites lived mainly in shallow waters and fed, in most cases, on what they could find in the sediment on the seafloor. Spectacular pyritized trilobites are found in the Hunsrück Slate, some specimens showing very fine details (Figure 64). The abundance of trilobites and the traces they produced show that they lived on the muddy Hunsrück seabed and were not transported from somewhere else. The preservation of trackways represents times when bottom currents were minimal, for only then can such evidence survive. Increased water movement, in contrast, may erode such trackways, introducing slightly coarser sediment and generating characteristic sedimentary structures.

Figure 63. The trilobite *Chotecops ferdinandi*. In this case the early formation of calcium phosphate around the decaying soft tissues during fossilization prevented the specimen becoming flattened (length of specimen 6.3 centimeters).

Figure 64. Detail of the rear limbs of the specimen of *Chotecops* sp. shown in Figure 66 (the area is 3.5 centimeters long).

Chotecops

A remarkably preserved specimen of *Chotecops* was buried lying on its side (Figure 65). During collection of the specimen from the spoil tip at the slate quarry, part of the thoracic skeleton was lost. This loss, however, turned out to be a bonus: it revealed the limbs hidden below. The two-branched limbs with their seven-segmented walking leg and filamentous gill branch are clearly evident, with even the finest details visible to the naked eye. Another special feature of this specimen is the three-dimensional preservation of the compound eye, which is essentially intact. This fossil required very careful and time-consuming preparation by Christoph Bartels, as well as the skills of an experienced photographer to obtain the best image. A second *Chotecops* that Bartels also prepared reveals the details of the ventral side (Figure 66). Specimens like this are important because they show extraordinary details of features such as the limbs, helping us to interpret the way the arthropod lived.

Six Specimens of the Trilobite *Chotecops*

The trilobite *Chotecops* was one of the most common inhabitants of the Hunsrück sea. Several specimens are often found on a single slate slab. Nevertheless, one such slab is unique because all six individuals on it have well-preserved limbs, including details of the walking legs and gill branches (Figure 67). The specimens show almost all the possible attitudes to bedding, providing a virtually three-dimensional perspective of the animal. Apart from the specimens already exposed on the surface, numerous other fossils remain to be prepared. X-radiographs show within the slab more than ten wormlike organisms of unknown nature about 2–3 centimeters in length, and countless other small objects that have yet to be interpreted.

The different positions of the trilobites relative to the bedding surface is of particular interest. Two specimens, one of them partly enrolled, show the dorsal surface of the exoskeleton. X-radiographs show that all the concealed limbs are preserved. An incomplete specimen at the edge of the slab is exposed from the ventral side. Another specimen provides a ventral view of the head, while the thorax is twisted along its length so that the rear part of the trilobite is exposed from the dorsal side. Two other specimens provide an almost precisely lateral view. As these specimens were flattened lying on their side, their bodies were compacted to about a quarter of their original thickness as the mud was transformed to slate. Even

Figure 65. Three views of a specimen of the trilobite *Chotecops* sp. Top, complete specimen (length 6.5 centimeters); middle, the rear part of the body with exposed appendages; bottom, the head showing the compound eye.

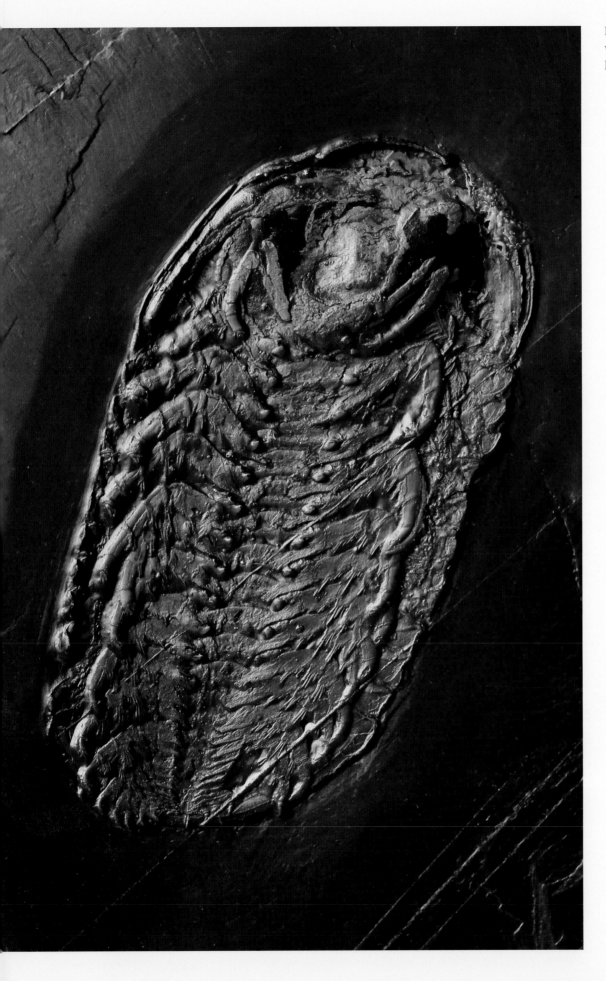

Figure 66. *Chotecops* sp. showing the ventral side with remarkably preserved limbs (length 8.5 centimeters).

though such a lateral orientation is not stable, these specimens remained in this position because of the cohesive nature of the sediment. Only transport and deposition in a turbulent cloud of sediment can explain the different orientations of the specimens: the slab represents the final stage of a turbidity current of considerable density. The laterally compacted specimen at the lower left, for example, is buried with the lower half of the body and the limbs of the right side in a layer of fine sand with a high content of pyrite. This sand is the coarser sediment that settled on the seabed first as the turbidity current waned. It was followed by finer muddy sediment that buried the rest of the animal. This single event deposited a layer that was eventually compacted to a thickness of slate of about 2 centimeters.

Figure 67. Slate slab (width 35 centimeters) with six specimens of the trilobite *Chotecops*. The different positions of the animals testifies to their deposition from a turbulent cloud of sediment.

Parahomalonotus

Trilobites, similar to present-day wood lice or pill bugs, were able to roll up in response to danger. Such a strategy would obviously have protected *Parahomalonotus* against an imminent threat (Figure 68). Indeed, this individual may have rolled up in response to a sudden influx of sediment, even though it was buried nonetheless. This specimen is from the roof slates of the Mayen region, which are still regarded as Hunsrück Slate even though they may be somewhat older than the sequence in the central Hunsrück region. The age of the Mayen slates is not precisely known, but their nature and depositional setting are essentially the same as those of the middle part of the Hunsrück sequence. An interesting feature of this fossil is the large cubic pyrite crystals on its surface. These formed sometime after the fossilization of the trilobite.

Kettneraspis

Like the specimen of *Parahomalonotus* displayed in Figure 68, the small trilobite *Kettneraspis* is enrolled (Figure 69). In this case, however, the front of the cephalon (head), which looks like a face, is turned toward the observer. The pygidium (tail) is curled beneath it, and its outline, with the front edge of the cephalon, forms a triangle. The glabella, an elongate, raised structure that housed the stomach, appears like a knobby nose in the center of the cephalon. A notable feature of this trilobite is the long, slender spines, which are typical of odontopleurids, the family to which *Kettneraspis* belongs. A long spine projects from the rear corner of the cephalon, and long slender spines are also present on the thorax and pygidium, becoming shorter toward the rear of the trilobite. In the photograph these spines are obvious to the left and around the margin of the pygidium. They are revealed as a result of the orientation of the buried trilobite relative to the surface along which the slate has split. Other species of *Kettneraspis* are known from rocks of Silurian to Middle Devonian age from all over the world. In the Hunsrück Slate, however, this trilobite is extremely rare.

Rhenops

A striking feature of the trilobite *Rhenops* is the spines that project from the corners of the cephalon (they are known as genal spines). The appearance of specimens of *Rhenops* can vary significantly depending on their orientation in the slate after they were buried by a turbidity current, and due to mineral growth after fossilization. In the example shown here, which is exposed from the dorsal side, the genal spines look very wide (Figure 70). This is the result of extension due to tectonic forces. Stretching created gaps that were filled with white quartz, distorting the true dimensions of the trilobite.

Figure 68. Enrolled trilobite *Parahomalonotus planus* with very large pyrite crystals (specimen diameter 6 centimeters).

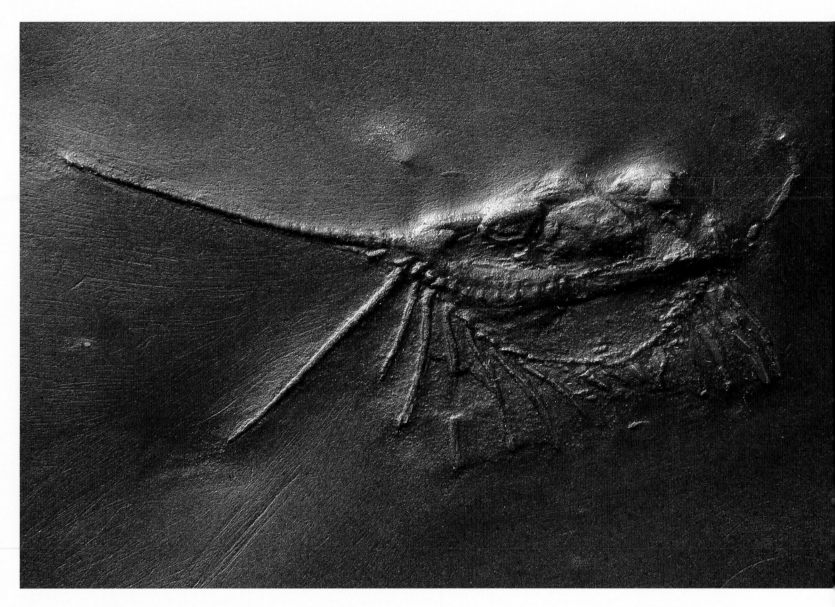

Figure 69. The trilobite *Kettneraspis* sp. showing the front of the cephalon,
behind which the body is enrolled (width 3 centimeters).

Figure 70. Dorsal side of a specimen of *Rhenops* sp. showing the growth of quartz around the genal spines (length of trilobite 6 centimeters).

Figure 71. A specimen of *Odontochile rhenanus* that measures 23 centimeters in length.

Figure 72. Detail of the compound eye of *Odontochile rhenanus* (the area illustrated is about 3.5 centimeters long).

Odontochile

In case you might be left with the impression that all the Hunsrück Slate trilobites were small, there is a specimen of *Odontochile* that measures at least 23 centimeters in length, although the triangular extension at the front of the cephalon has been lost (Figure 71). Imagine a woodlouse or pill bug this large!

Both the right and left antennae are evident, projecting from either side of the cephalon. The eyes are almost perfectly preserved: the arrangement of the individual lenses is highly symmetrical (Figure 72). Such complete specimens of this trilobite are extremely rare in the Hunsrück Slate.

Sea Spiders (Pycnogonida)

Evolution has sometimes resulted in remarkable forms of life. The pycnogonids, or sea spiders, are undoubtedly a case in point. These animals seem at first glance to consist only of legs (Figure 73). The four most obvious pairs of legs, which are used for walking, account for the general similarity to spiders. The body is so small and slender that it cannot accommodate all of the animal's organs, which in some cases extend into the limbs. The front section of the body has three additional pairs of limbs—claws for cutting the prey, limbs for manipulating food, and the so-called oviger, which is generally used for cleaning, or by the male to carry around the eggs until they develop into larvae. The head extends into a proboscis, which terminates in the mouth. Adult sea spiders are predators. They feed on snails, sponges, bryozoans, cnidarians, and other small soft-bodied organisms.

The first sea spider discovered in the Hunsrück Slate, *Palaeoisopus problematicus,* was described by Ferdinand Broili in 1928.[22] This Devonian sea spider has long been the first and oldest evidence that these arthropods lived in the Paleozoic seas. Meanwhile, examples have been found in the Silurian and possibly even in the Cambrian. Four species have been found so far in the Hunsrück Slate. An analysis of their relationships has shown that they are unlikely to be closely related to modern forms.

Palaeoisopus

This pycnogonid appears to confirm the idea that size *does* matter! The spread of the first pair of legs reaches up to 40 centimeters, making *Palaeoisopus problematicus* not only the largest sea spider in the Hunsrück sea, but also one of the largest Hunsrück Slate fossils. It was probably an important predator.

The most striking feature of *Palaeoisopus problematicus* is the huge first pair of walking limbs, which extend to the upper corners of the photograph (Figure 74). These limbs end in large, robust, movable spines that were probably used for gripping prey and climbing, on crinoid stems, for example. The left and right limbs appear different in length and shape because they lie in different orientations in the slate. The short head limbs used to manipulate the food are preserved projecting down between the first pair of large walking limbs in this specimen. The second walking limb on the left side appears thin and straight compared with the others. Here the underside of the limb is

Figure 73. Alcohol preparation of a recent sea spider, *Boreonymphon abyssorum,* from the Arctic Sea (diameter 4 centimeters in the stance depicted). Collection of the Zoological Museum, University of Göttingen.

exposed, in contrast to the other limbs, which were buried in such a way as to afford a lateral view. These limbs were clearly flattened in life and may also have been used for swimming. Even the series of fine bristles that project from their inner margin is evident.

Another specimen of *Palaeoisopus problematicus* is preserved with the first pair of limbs projecting forward, toward the top of the image (Figure 75). The segments of the walking limbs with their fine bristles are very well preserved, the exoskeleton appearing almost like a suit of armor. The narrow abdomen with five divisions that is clearly displayed in this specimen is reduced to a short projection in today's sea spiders.

Bundenbachiellus

The arthropod *Bundenbachiellus giganteus* is known from just two specimens. One individual, which is essentially complete, was found in the 1990s (Figure 76). It shows the short head and long trunk with branched limbs, the lower branch stout and segmented, the upper consisting of a series of filaments.

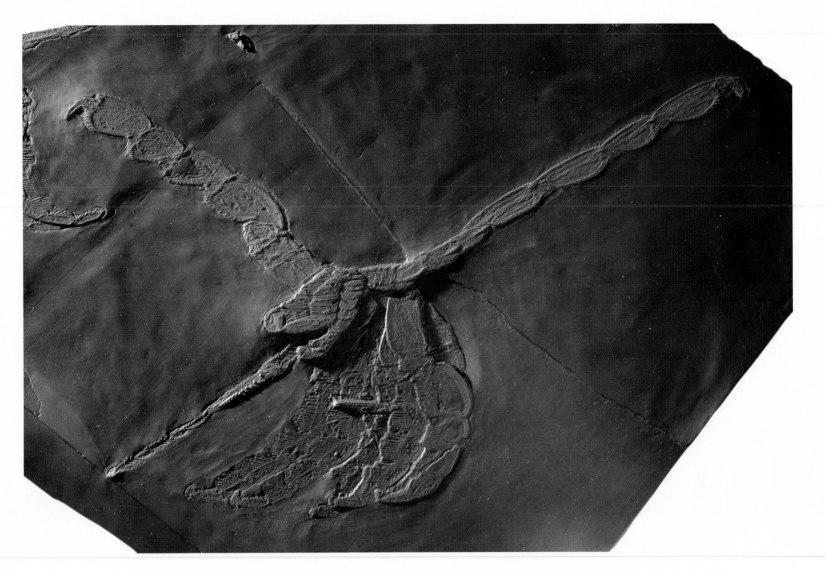

Figure 74. The sea spider *Palaeoisopus problematicus* (limb span 40 centimeters).

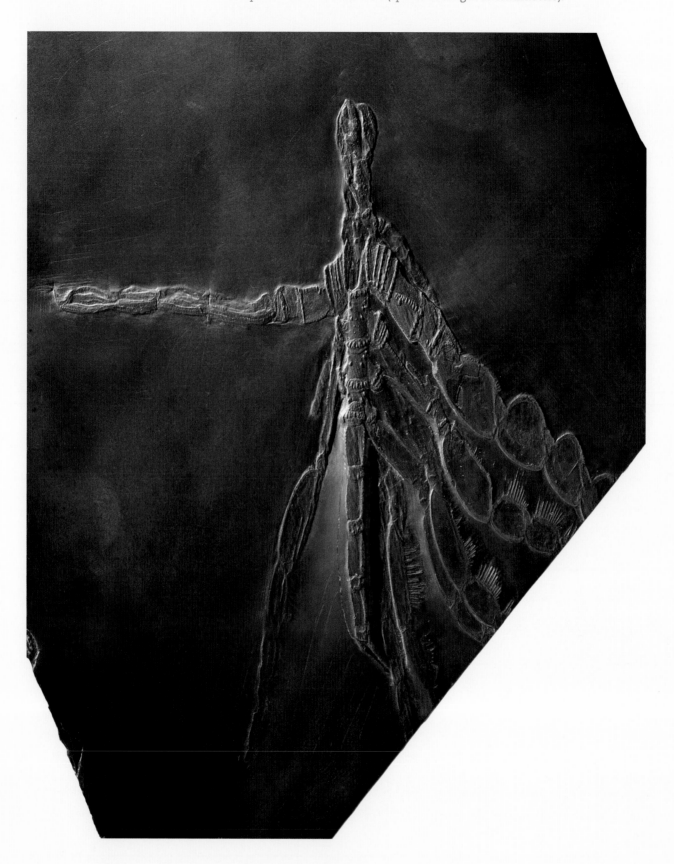

Figure 75. A second specimen of *Palaeoisopus problematicus* with the first pair of limbs outstretched (specimen length 22 centimeters).

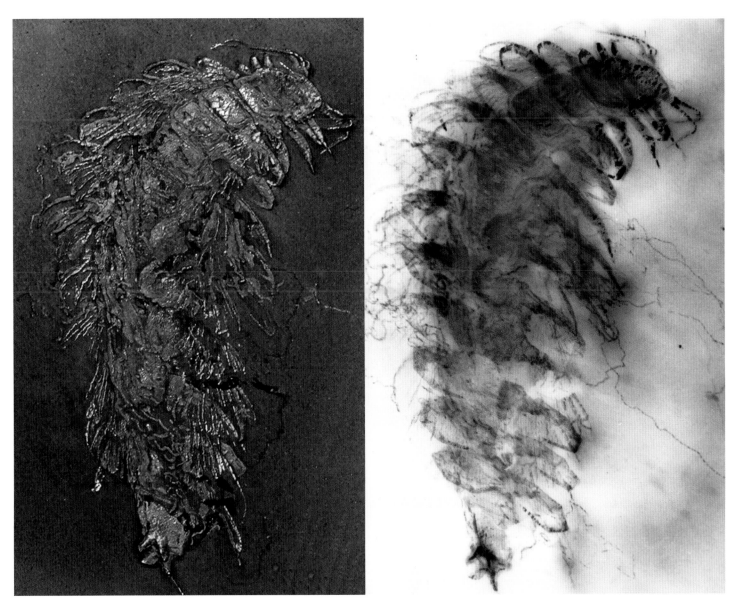

Figure 76. *Bundenbachiellus giganteus*, with an x-radiograph of the fossil (length 6.3 centimeters).

But *Bundenbachiellus giganteus* has had a checkered history. Ferdinand Broili described the only other specimen, which is about three and a half times longer, as *Megadactylus giganteus* in 1929.[23] Unfortunately the name *Megadactylus* had already been used for a dinosaur, so Broili gave the arthropod a new name, *Bundenbachiellus*, in a paper published the following year. This first specimen lacks the head, which made it difficult to interpret, and in 1978 it was identifed as a distorted specimen of the arthropod *Cheloniellon calmani* (see Figure 124).[24] With *Bundenbachiellus giganteus* "concealed" as a specimen of *Cheloniellon*, when the second specimen was discovered, it was thought to be a new arthropod and was named *Eschenbachiellus wuttkensis*.[25] It was only when Broili's specimen of *Bundenbachiellus giganteus* was reexamined in the Bavarian State Collection in Munich that it became clear that it and *Eschenbachiellus wuttkensis* are the same arthropod; the original name has priority.[26]

The relationships of *Bundenbachiellus* are uncertain, but it probably branched off the arthropod tree before the ancestors of the living groups. Like *Mimetaster*, it may be related to Cambrian forms.

The Sea Urchin (Echinoidea)
Rhenechinus

Unfortunately, one edge of this extremely rare fossil *Rhenechinus hopstaetteri* (only three specimens are known) was cut off when the block of slate containing it was trimmed, even before the specimen was discovered (Figure 77). So part of the sea urchin was lost before the slate was split. The specimen consists of the nearly spherical skeleton, or *test,* which consists of rows of hexagonal plates. Five bands of plates, running from the top of the sea urchin to the mouth on the ventral side, carried tentacle-like structures called tube feet, which are extensions of vessels inside the test that make up the water vascular system (see Figure 88). Although the tube feet are not preserved, the position of each one is marked by a pair of pores passing through the plate. The tube feet could attach to surfaces and were used by the echinoid for respiration. The five bands carrying the tube feet were separated by wider areas, which were also covered by hexagonal calcareous plates. The plates of the test were not fused together, unlike those in most modern sea urchins, but connected by soft tissue, making the test flexible. Flexible tests are known today only in the so-called leather urchins (echinothurioid echinoids), such as *Phormosoma bursarium* (Figure 78). *Rhenechinus hopstaetteri* bore numerous slender spines, finely ribbed along their length, which are evident in Figure 77. They were most densely packed around the tube feet in order to protect them. Inside the test, and evident only in an x-radiograph of *Rhenechinus,* is the jaw, an elaborate structure known as Aristotle's lantern (Figures 79 and 80).

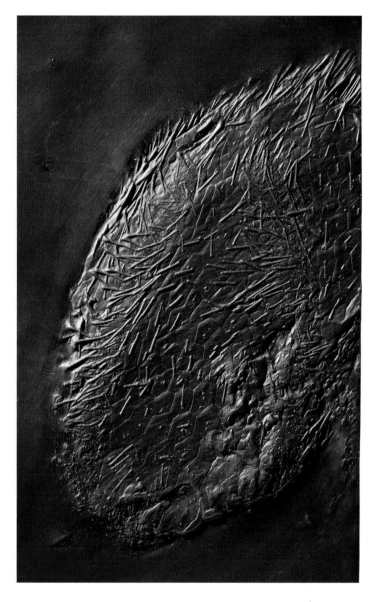

Figure 77. The sea urchin *Rhenechinus hopstaetteri* (diameter 12 centimeters), part of which was sawn off during slate processing.

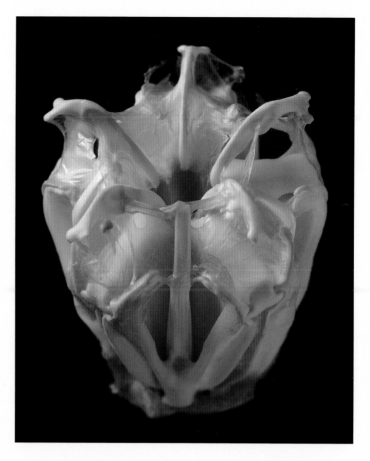

Figure 78. Completely flattened example of the modern leather urchin *Phormosoma bursarium* from Balicasag Island, Philippines (diameter 8.5 centimeters). Only the Aristotle's lantern, the jaw of the echinoid, shows much relief. Collection of Volker Thiel, Düsseldorf.

Figure 79. Aristotle's lantern of the sea urchin *Diadema antillarum*, from the east coast of the United States (diameter 2.3 centimeters). Collection of Volker Thiel, Düsseldorf.

Figure 80. The lower half of the test of the sea urchin *Sphaerechinus granularis* (diameter 9 centimeters), exposing the Aristotle's lantern, from the Mediterranean Sea. Collection of Katharina Händeler, Bonn.

Figure 81. Large slate slab with numerous specimens of the brittle star *Furcaster palaeozoicus* (size of the slab 40 by 80 centimeters).

Brittle Stars (Ophiuroidea) and Sea Stars (Asteroidea)

Who is not familiar with these animals? Anyone who has vacationed at the seaside even once has probably seen a sea star, if nowhere else perhaps as a dried specimen in a souvenir shop. Brittle stars are less well known, although they look very similar to sea stars.

These groups are represented in the Hunsrück Slate by a total of about fifty species. Although some species are very rare, others occur relatively frequently. We can usually identify these animals only when their skeletons remain more or less intact, because the calcareous elements that make up the skeleton are very small and difficult to distinguish when isolated.

Brittle Stars (Ophiuroidea)

Brittle stars differ from sea stars in their clearly defined body, which normally bears five long, slender snakelike arms. Brittle stars are the most diverse group of echinoderms today, with more than two thousand species. They are relatively common in the Hunsrück Slate, which is among the world's most important sources of fossil ophiuroids. Occasionally well over one hundred specimens are preserved on a single piece of slate the size of a small table, representing a mass mortality (Figure 81). Mass occurrences of brittle stars and other animals, and the large size of specimens relative to similar fossils from other sites, indicate that living conditions in the Hunsrück sea were unusually favorable, including a rich food supply. Slate mining has revealed mass occurrences covering several square meters.

Nurseries of brittle stars, where juveniles congregated, are also found in the Hunsrück Slate. All of the common species are represented by a range of sizes from tiny juveniles to large adult specimens (sometimes up to half a meter in diameter).

Bundenbachia

This brittle star is one of the more common and characteristic fossils of the central Hunsrück region (Figure 82). The arms are lanceolate in shape, with long, whiplike terminations. Relatively short, strong spines are present along the arms.

Loriolaster

The spectacular specimen here of *Loriolaster,* a brittle star named after the nineteenth-century fossil echinoderm expert Perceval de Loriol, preserves the skin stretched between the arms in such detail that even the finest creases are still visible (Figure 83). The specimen is exposed from the dorsal side. Even the lateral spines on the ventral side of the arms are evident under the covering of skin. The change in color between the golden pyrite of the skin and the dark color of the arm plates shows that different chemical processes were involved in their preservation. Pyrite precipitated on the organic skin, but the plates required replacement of the original calcium carbonate.

One large slab of slate preserves a mass mortality of *Loriolaster* and other brittle stars (Figure 84). Numerous animals were transported together and buried, and it is interesting to note that different specimens show different states of preservation, even on the same slab. Some are replicated in pyrite. This process, however, affects only certain parts, particularly the skin. Blackish specimens, in contrast, are at least partly preserved in the calcium carbonate of the skeleton, and their soft tissues have decayed further. These differences in the style of preservation may reflect small-scale variations in the sediment or the susceptibility of the tissues to decay. Even though there are signs that these brittle stars were brought together during sediment transport, this group probably reflects dense settlement of areas of the seafloor by just one or two types of brittle star. Such concentrations are also known today.

Figure 82. The brittle star *Bundenbachia beneckei* (diameter 10 centimeters).

Figure 83. The brittle star *Loriolaster mirabilis* (diameter 12 centimeters).

Furcaster

Furcaster is the most common brittle star in the Hunsrück Slate (Figures 85 and 86). These animals are characterized by slender arms with delicate spines and a small body disk that was probably originally convex, with small calcareous plates in the skin. After the animal died the body disk, which accommodated the digestive organs, collapsed and was often distorted during compaction. Many specimens show early signs of decay including bursting of the disk. Although the center of the body may be distorted, it is evident where the term "ophiuroid," the scientific name for brittle stars, comes from—it means "snake tail" and refers to the appearance of the arms. Not only does the animal as a whole form a five-armed star, but the plates that frame the mouth in the center of the body disk also form a distinctive five-pointed star. This frame incorporates the first plates of the arms, which are the largest single elements in the skeleton.

Euzonosoma

Although the specimen of *Euzonosoma* pictured here is not very attractive, it is important for scientific study (Figure 87). This brittle star died some time before it was buried. The disintegration of the animal is clearly visible, and not much of the body remains. What survives, however, is important evidence of the process of decomposition, and the specimen provides clues to the identification of even less well preserved examples. In fact, only a small proportion of life in the Hunsrück sea is well preserved. Most individuals decayed in the normal way, and are represented, if at all, by fragments of the skeleton. The sedimentation events that resulted in the well-preserved fossils were rare exceptions in this environment, not everyday occurrences. Such unusual events, however, were more frequent in the layers that are exposed at the main localities around Bundenbach in the central Hunsrück region than elsewhere.

Figure 84. Slate slab (height 42 centimeters) with many specimens of *Loriolaster mirabilis*.

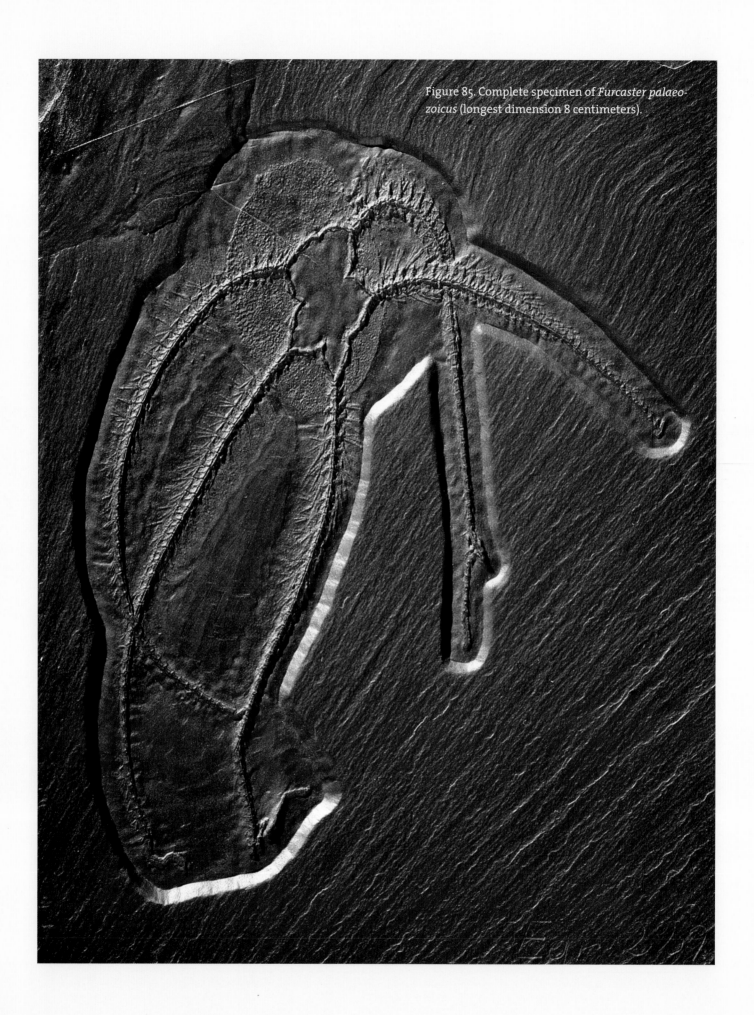

Figure 85. Complete specimen of *Furcaster palaeozoicus* (longest dimension 8 centimeters).

Figure 86. *Furcaster decheni* with four complete arms, and the fifth cut short at the edge of the slate slab (longest dimension 13 centimeters).

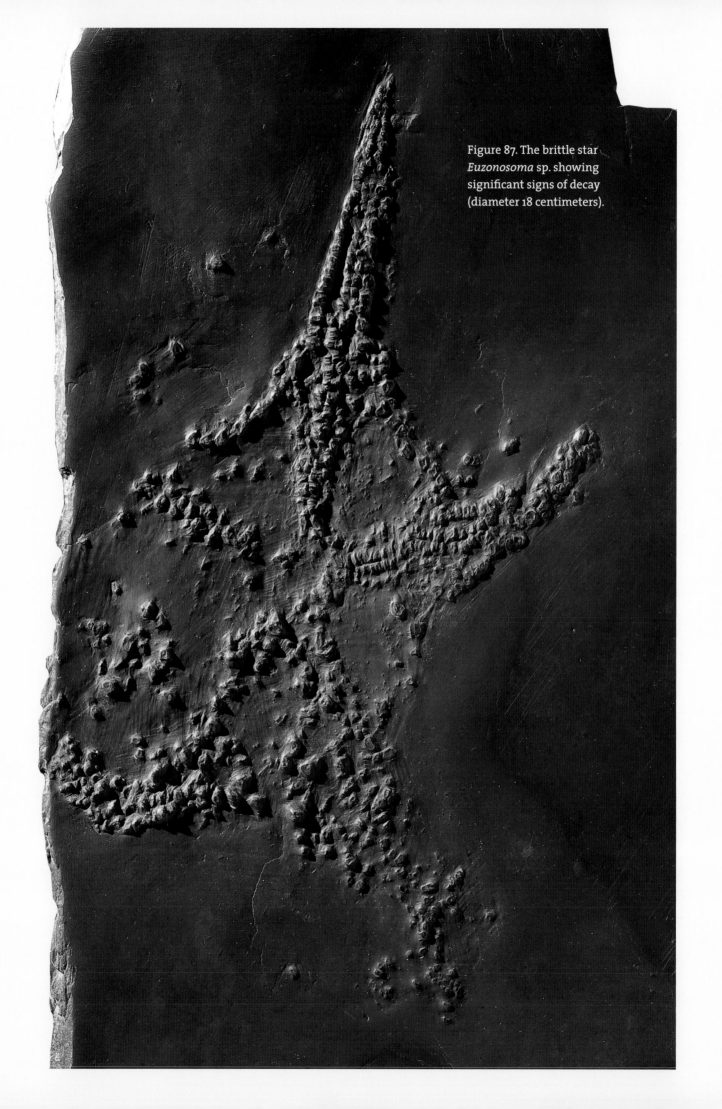

Figure 87. The brittle star *Euzonosoma* sp. showing significant signs of decay (diameter 18 centimeters).

Sea Stars (Asteroidea)

Although they are not as common as brittle stars, sea stars occur in considerable numbers and a large variety of forms in the Hunsrück Slate. Their characteristic body shape shows little difference, at least in general terms, from that of today's sea stars. It is not clear, however, to what extent their diet was also similar. Some of today's sea stars are predatory, and they can evert most of the stomach to digest prey externally, managing even large clams. Only unusual forms (some members of the Paxillosida) live on soft sediment and ingest shells or sediment through a highly flexible mouth opening. It is striking that many of the Devonian forms also had a relatively large mouth aperture; they may have fed in a similar way.

The arms of sea stars bore numerous movable spines along their margins. Particularly spectacular are specimens of the many-armed "sun star" *Helianthaster,* which are extremely rare. Sea stars normally have five arms, but *Helianthaster* has many more than that. Modern five-armed forms sometimes occur with four or six arms, even though they are otherwise perfectly normal. Sea stars can sometimes regenerate arms that have been partly or completely lost. This striking regenerative capacity also enables some sea stars to multiply asexually,

and some living species of the sea star *Linckia* can regenerate a complete specimen from a severed arm.

Sea stars move with the help of their tube feet, extensions of the internal network of connected vessels, called the water vascular system, that is a feature of all echinoderms. Five radial canals extend into the arms of a sea star and around the test of a sea urchin (Figure 88). Small tube feet with muscular sacs branch from these radial canals in pairs at regular intervals. The tube feet can be extended or retracted by changing the volume of the muscular sacs. The tube feet are useful in many ways, including feeding, locomotion, respiration, and sensing chemical signals in the water.

Tube feet are extremely delicate and decompose rapidly after the death of the animal. They have yet to be found in a Hunsrück Slate sea star, but the discovery in 2004 of tube feet preserved in a brittle star caused a sensation.[27] They were found in specimens of *Bundenbachia beneckei* by careful preparation using an air-abrasive machine with iron powder (Figure 89). This method may reveal preserved tube feet in other Hunsrück Slate echinoderms in the future.

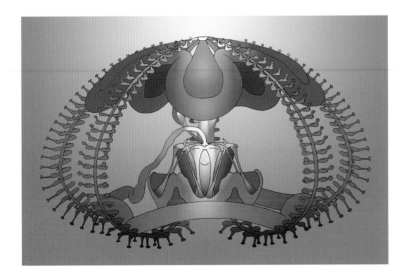

Figure 88. Diagram of the interior of a sea urchin, including the water vascular system. Sea stars, brittle stars, and crinoids have similar systems.

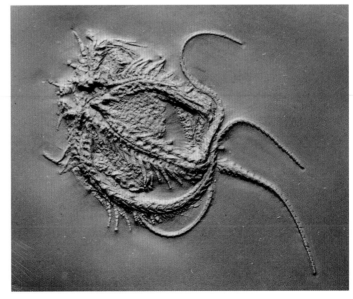

Figure 89. The brittle star *Bundenbachia beneckei,* with preserved tube feet (diameter 8 centimeters).

Urasterella

Urasterella asperula is one of the most common sea stars found in the Hunsrück Slate in the central Hunsrück region. The example pictured here, which shows the under side, features the small calcareous plates of the arm arranged in uniform rows, and the slender marginal spines are evident (Figure 90). The sea star grew by adding plates to the end of the arms. The plates were connected by small muscles and connective tissue and covered by the skin. The grooves that run along the under side of the arms are clearly visible in this specimen. Food particles were transported along these grooves to the mouth in the center of the body. Specimens with arms ranging from 0.5 centimeter to nearly 15 centimeters in length have been found, representing all the growth stages.

Baliactis

Baliactis tuberatus is among the rarest fossils in the Hunsrück Slate: we are aware of only five examples. The specimen shown here is one of the most spectacular sea stars ever found in the Hunsrück Slate (Figure 91). The grooves that run along the middle of the arms on the under side are clearly evident. They are flanked by calcareous plates with a series of small pits. These pits accommodated the tube feet in the living animal, but the tube feet are not preserved here. Large plates on the upper side of the arms were covered in tubercles. The arm at the top of the photograph was twisted during burial, so that part of the upper side is showing here.

Figure 90. *Urasterella asperula* (diameter 9 centimeters).

Figure 91. *Baliactis tuberatus* (diameter 7 centimeters).

Helianthaster

This spectacular, many-armed sea star, popularly known as a "sun star," usually has sixteen arms (Figure 92). The mouth frame looks like a rosette. The x-radiograph clearly shows the construction of the arm skeleton (Figure 93). It consists of small calcareous plates arranged symmetrically on both sides of the arms, from which small spines project. The body disk, with big robust plates framing the mouth opening, was covered by a skin containing granular calcareous elements. These elements increase in size and density toward the edge of the body disk, as shown in the x-radiograph. The elements of the skeleton were replaced by pyrite, which absorbs x-rays. Thus they show up in an x-radiograph as white or gray areas, faithfully tracing the skeleton of the sea star. When this x-radiograph is printed, as here, the skeleton appears dark.

The first x-radiographs of *Helianthaster* were made by Walther Lehmann (1880–1959). Following an early career in business, he changed direction and received his Ph.D. in physics in 1924, at the age of forty-four. The Hamburg-based paleontologist Georg Gürich suggested to Lehmann that he should try applying x-ray techniques to fossils. With a specially designed x-ray laboratory, Lehmann settled in Kirn and began to study the fossils from the Hunsrück Slate. In 1941 he was appointed a lecturer in petrology and geology at the University of Bonn, where he took up a professorship in 1946. Lehmann's 1957 monograph on fossil sea stars and brittle stars from the Hunsrück Slate helped to put both the Hunsrück region and the application of x-rays to fossils in the limelight.[28] After Lehmann died it was generally thought that the Hunsrück Slate fossils had been exhaustively researched and published. Roof slate mining was soon brought to a virtual standstill, so that no new discoveries were being made. The revival of slate mining in the 1970s to the 1990s revealed that new discoveries could be made about the fossils of the region. Remarkable new specimens of sea stars have been found, like the example depicted in Figure 93. Perfectly preserved specimens like this, which have been collected intact, are still rare enough to strike us as extraordinary.

Figure 92. Lithograph of *Helianthaster rhenanus,* from Roemer (1863), one of the earliest publications on the fossils of the Hunsrück Slate.

Palaeostella

The x-radiograph of this compact sea star demonstrates the possibilities of this imaging technique for studying the Hunsrück Slate fossils (Figure 94). Two arms of the sea star are folded on themselves as a result of transport in the sediment-laden current that buried this individual. The central mouth region shows signs of early decay. The x-rays reveal evidence of other activities in the mud. Stringlike structures in the slate around the fossil, some of which appear to branch, are often made up of small discrete elements preserved like a string of beads. These are likely the pyritized fecal pellets of small, wormlike sediment feeders, left in the mud as the worm moved forward, and colonized by microbes. These microbes generated conditions for the formation of pyrite. The appearance of branching may be deceptive; it is sometimes created by two such strings of pyrite overlying each other at different levels in the slate slab, which are superimposed in the x-radiograph.

Hystrigaster

Hystrigaster horridus was first discovered and described by Walther Lehmann in 1957 (Figure 95). The name reflects the peculiar morphology of this fossil sea star, which prompted an otherwise serious scientist to make a somewhat fanciful comparison! The long, stout, movable spines reminded Lehmann of the quills of a porcupine, and he interpreted them as a defense against possible predators. The scientific name he coined means "dreadful porcupine star." The large, sturdy arms of *Hystrigaster* have robust plates. The convex body disk is also covered by a network of plates, with small tubercles where they meet. Each of these tubercles bears a large, movable spine. Rows of spines are also present on the arms. The mouth region of *Hystrigaster* is unusually large, with robust mouth-angle plates. Traces of movement by the sea star are preserved on the slate slab. It was presumably buried by a sediment cloud carried by a turbidity current, which clogged its water vascular system and led to death. A specimen of the brittle star *Loriolaster* is preserved on the same slab but in a different layer of sediment, at a level approximately 1.5 centimeters above the *Hystrigaster* (other examples of *Loriolaster* are portrayed

Figure 94. X-radiograph of *Palaeostella solida* (diameter 10 centimeters).

Figure 95. The sea star *Hystrigaster horridus* together with a specimen of *Loriolaster* (the area illustrated is 20 by 30 centimeters).

in Figures 83 and 84). This *Loriolaster* was not buried by the same sedimentation event as the *Hystrigaster*. Its appearance suggests that it was already dead when it was picked up by a mud-laden current, its muscles passive and arms floppy. Similar specimens show different stages in the decomposition of the skin between the arms.

Protasteracanthion

Protasteracanthion primus is one of the rarest fossils in the Hunsrück Slate (Figure 96). It is not yet clear whether this animal belongs with the sea stars or the brittle stars. Specialists sometimes find such questions difficult to resolve based on the evidence of fossils, even in examples of exceptional preservation like those from the Hunsrück Slate. The body disk of

Protasteracanthion is very small, and the long, slender arms suggest that the animal could move quickly by wriggling them. The delicate nature of the skeleton, with its regular arrangement of many arm plates that become smaller and smaller toward the tips, is well illustrated by this x-ray image. A small piece of one of the arms has been lost. The arms look like very finely crafted bracelets. In some parts of the skeleton, however, there are coarser, more opaque structures. These are large crystals of pyrite that obscure details of the skeleton where they have grown along parts of the arms. These differences are often the result of deformation of the rock during the tectonic events that happened later. As the slate formed, tectonic forces caused stretching and cracking; the spaces that were created were filled partly with pyrite and partly with quartz.

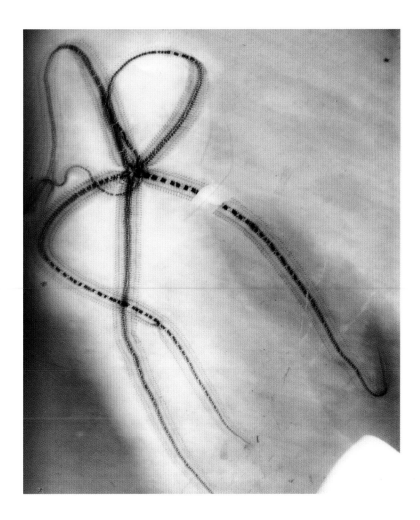

Figure 96. X-radiograph of *Protasteracanthion primus* (greatest diameter 16 centimeters).

FREE SWIMMERS

Ammonites

Arthropods

Fishes

Ammonites (Ammonoidea)

Ivoites is an early representative of the ammonoids, a very successful group of cephalopods (animals such as octopuses, squids, cuttlefish, and *Nautilus*) that includes the familiar ammonites of the Jurassic and Cretaceous. The body, similar to that of today's *Nautilus,* was largely concealed in a chambered shell (Figure 97). Ammonoids were once very common in all the world's oceans, before they became extinct at the end of the Cretaceous. Devonian examples, such as *Ivoites,* are among the oldest representatives of these animals, but they are relatively rare in the Hunsrück Slate. This rarity is probably a result of their free-swimming mode of life, which helped them to avoid becoming caught up in turbidity currents and buried. There are more than ten species of ammonoid in the Hunsrück Slate. Their diet is unknown, but they may have fed on smaller creatures such as arthropods and other mollusks. They themselves were probably a source of food for larger predators. The specimen of *Ivoites* illustrated here preserves traces that were previously interpreted as a bite mark made by an unknown attacker (Figure 98). Other Hunsrück Slate specimens, however, have up to eight indentations of this kind, which means that they are unlikely to be the result of a bite. These structures have been interpreted more recently as a result of the activity of some unknown parasite. In contrast to other fossil-bearing Hunsrück Slate slabs, the surface surrounding this specimen of *Ivoites* is uneven. Close inspection reveals the presence of many tentaculitids, the tiny cone-shaped shells that are sometimes found adhering to other animals.

Figure 97. *Ivoites* from the Hunsrück Slate with three marks on the body chamber (diameter 11 centimeters).

Figure 98. Detail of one of the marks on *Ivoites* (area illustrated approximately 3 centimeters wide).

Arthropods (Arthropoda)

Schinderhannes

The discovery of *Schinderhannes bartelsi,* described for the first time in 2009 by Gabriele Kühl and her colleagues, was a scientific sensation.[29] This unusual predator is like no animal living today (Figures 99 and 100). It was named after a famous eighteenth-century robber of the Hunsrück region, Johannes Bückler, who was known as Schinderhannes. The head of *Schinderhannes bartelsi* bears a pair of very large anterior grasping limbs, which were used to capture prey. These grasping limbs, a round mouth lined with small plates, and very large, stalked compound eyes represent a combination of head features that is known only in animals like *Anomalocaris* and *Hurdia* from Cambrian marine deposits some 100 million years older, such as the Burgess Shale. The body, on the other hand, has the series of dorsal plates (called tergites) and two-branched limbs characteristic of arthropods. Thus *Schinderhannes* combines features of *Anomalocaris* with those of early true arthropods. Until recently it was believed that anomalocaridids, the group comprising *Anomalocaris* and its close relatives, became extinct in the Cambrian. The discovery of *Schinderhannes* showed that features of anomalocaridids survived at least until the Early Devonian. This discovery raised the question of what related forms were present in the intervening 100 million years. As if on cue, a giant anomalocaridid was reported in 2011 from the Ordovician of Morocco.[30]

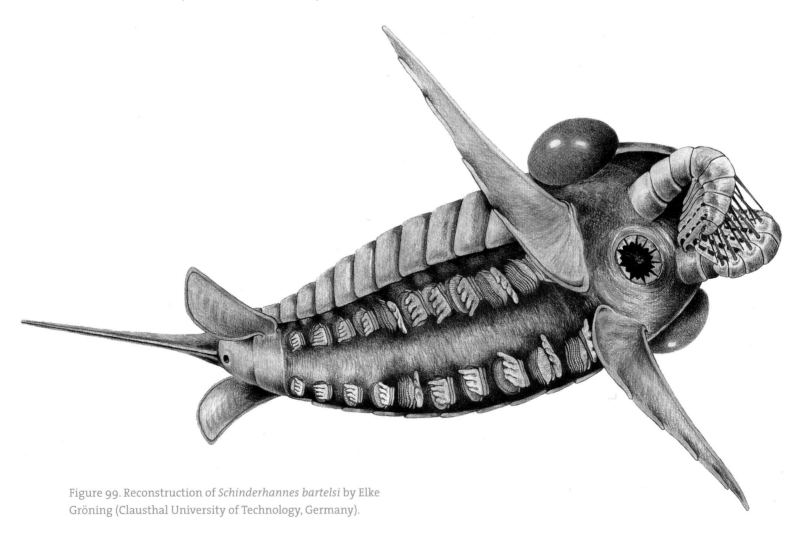

Figure 99. Reconstruction of *Schinderhannes bartelsi* by Elke Gröning (Clausthal University of Technology, Germany).

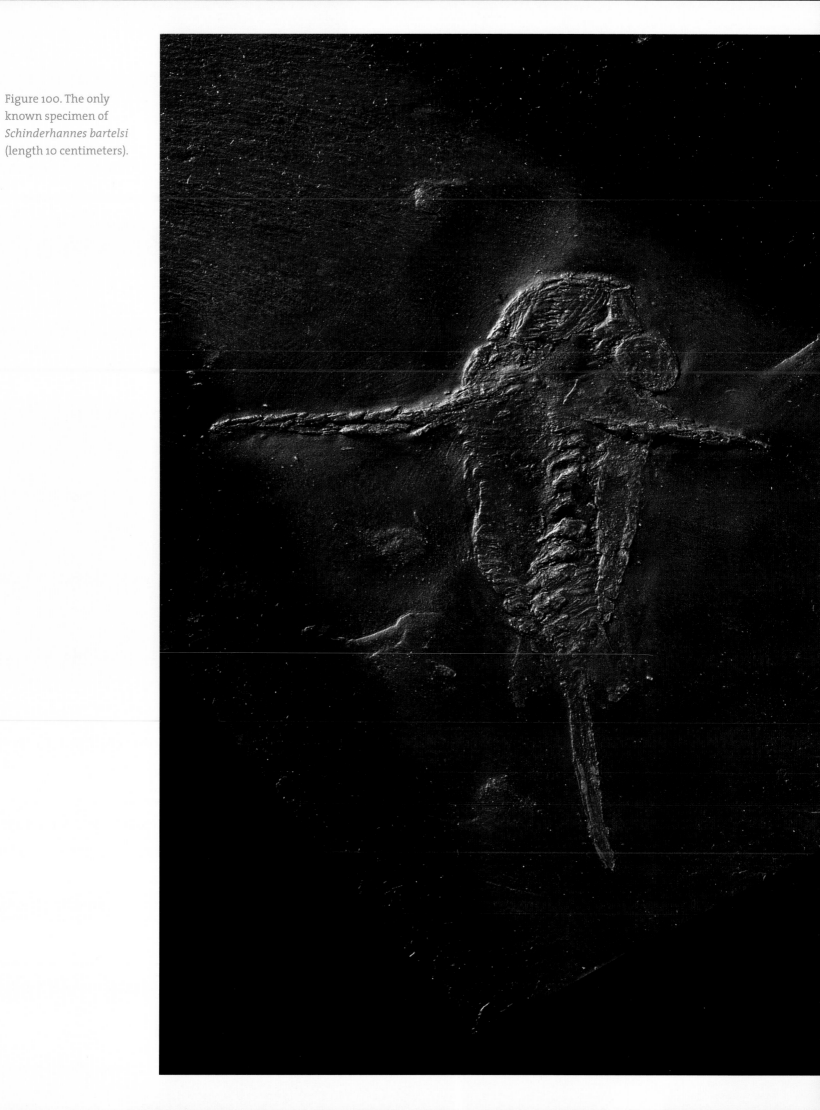

Figure 100. The only known specimen of *Schinderhannes bartelsi* (length 10 centimeters).

Crustaceans (Crustacea)

Crustaceans are represented in the Hunsrück Slate by at least seven different species, of which the phyllocarids are the best known. Three of these, *Nahecaris stuertzi, Nahecaris balssi,* and *Heroldina rhenana,* are the most extensively studied. The phyllocarids in the Hunsrück sea were veritable giants compared with their modern relatives, which are only a few centimeters in size (for example, *Nebalia,* Figure 101). *Heroldina rhenana,* in particular, reached a length of up to 60 centimeters, or roughly the size of some of today's larger lobster species.

Figure 101. The living phyllocarid *Nebalia;* these crustaceans reach sizes of only a few centimeters.

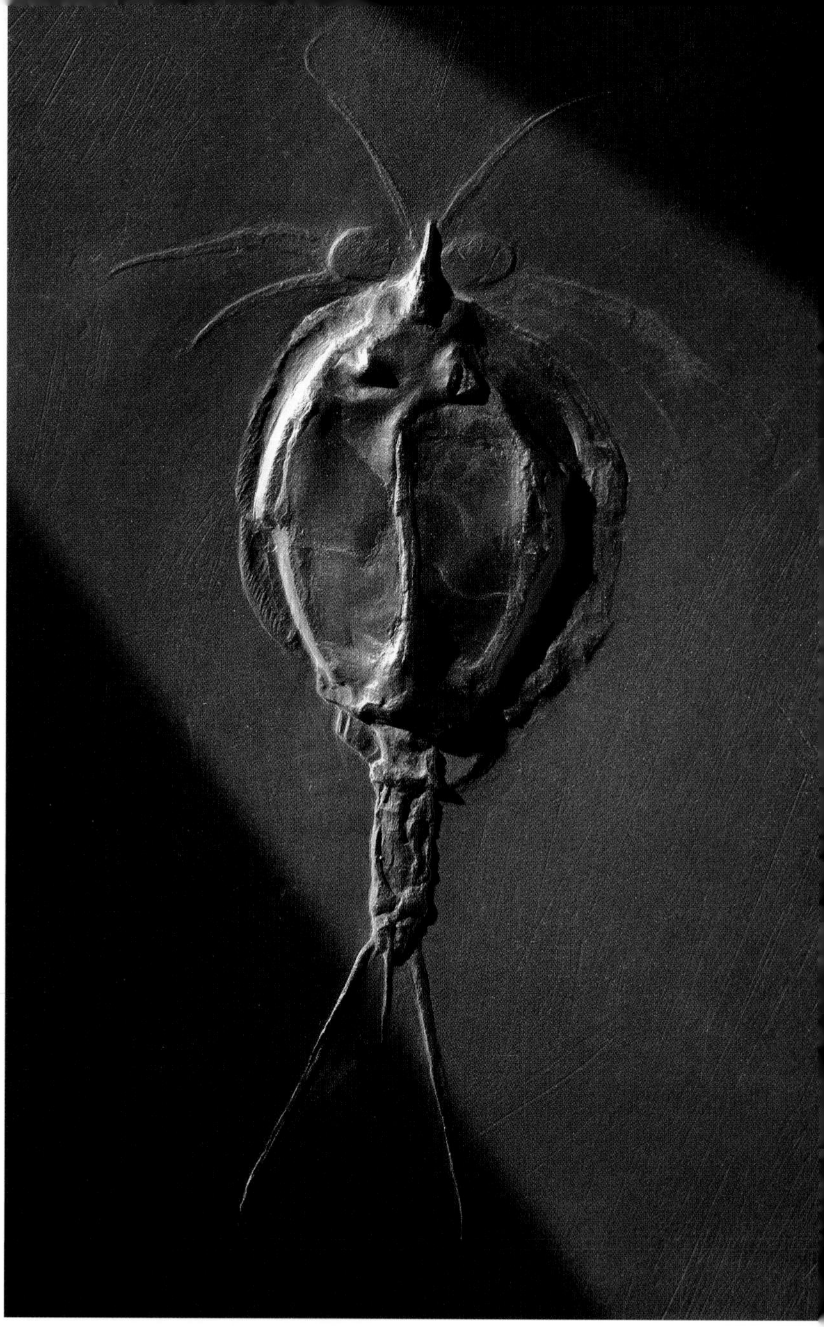

Figure 102. *Nahecaris stuertzi* (length about 10 centimeters).

Nahecaris

Nahecaris stuertzi was one of the more common crustaceans in the Hunsrück sea (Figures 102 and 103). The hinged carapace, with a spinelike projection at the front, and the long tail fork are the most striking features of this form. Reaching up to 17 centimeters in length, it was considerably larger than its present-day relatives. These animals lived on the substrate and used their limbs to stir up the mud in search of food. They were also able to swim using flaplike limbs on the abdomen, which extends beyond the carapace.

One extraordinary slate slab preserves a pair of individual specimens side by side, although whether these represent a male and a female is unknown (Figure 104). In any case, they were clearly together when they were buried unexpectedly. The carapace of the lower specimen is covered in a very thin brassy layer, the result of preparation with a brass brush, which gives it the appearance of being crafted from precious metal. Only the upper part of the two valves of the carapace is visible, separated by a very obvious hinge line. The lower halves of the valves are folded beneath and concealed in the sediment. The upper specimen, in contrast, provides a lateral view showing the full outline of the right valve. This difference is a consequence of the burial of the two individuals in different orientations in the mud deposited on the Hunsrück seafloor.

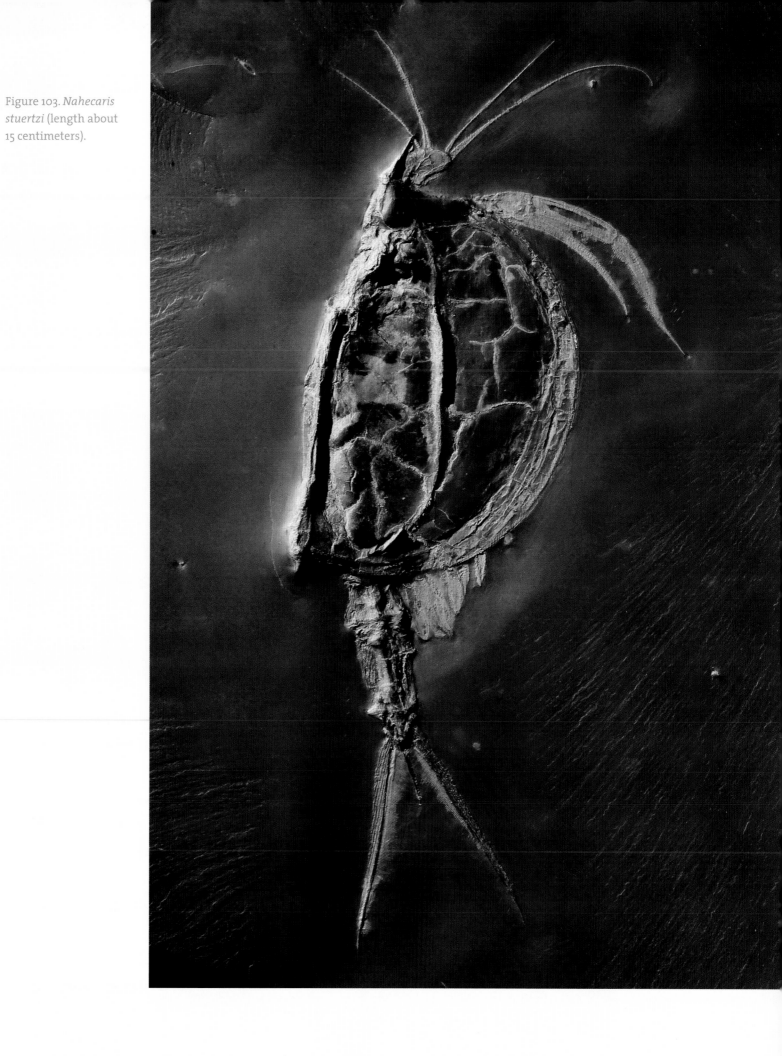

Figure 103. *Nahecaris stuertzi* (length about 15 centimeters).

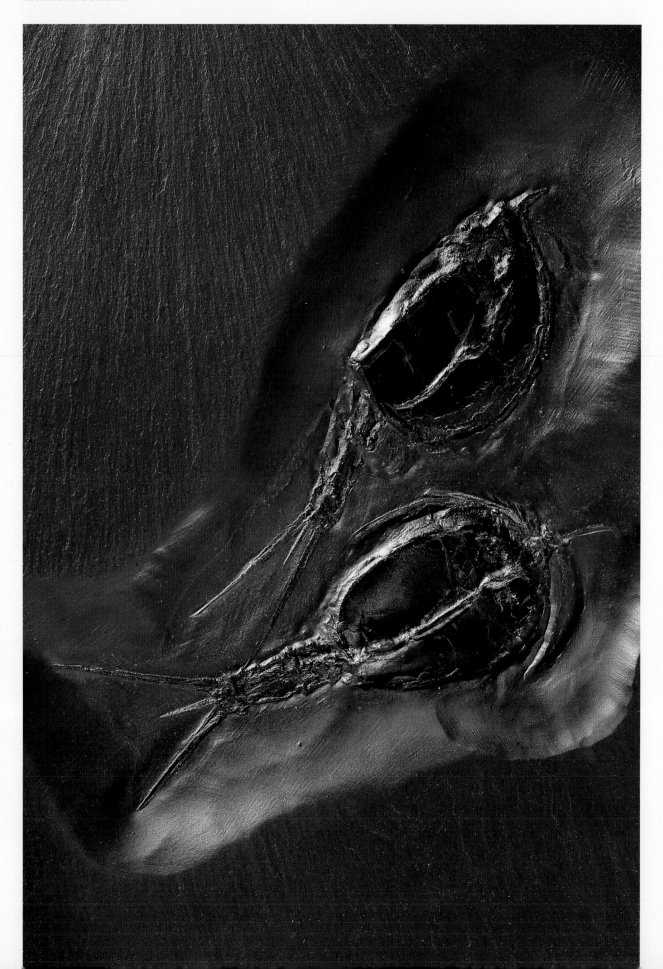

Figure 104. Two individuals (each 12 centimeters long) of *Nahecaris stuertzi*.

Nahecaris balssi is the smallest of the Hunsrück Slate phyllocarids; the specimen pictured here is about 5 centimeters long (Figure 105). Only about twenty specimens of this species are known. Almost all were discovered in a relatively short time in the upper layers of the last Hunsrück Slate quarry mined in the Bundenbach area. Marine life was not evenly distributed on the Hunsrück seafloor; certain species or groups of species were more prevalent in certain localities. The specimen in this photograph was prepared from the ventral side. One of the large compound eyes is clearly evident overlying the base of the first pair of antennae. Eight pairs of legs curve toward the midline of the trunk. They differ considerably from the swimming limbs of the abdomen behind. The middle spine of the tail (the telson) is about the same length as the two lateral spines (the telson limbs or tail fork), whereas it is much shorter in *Nahecaris stuertzi*. *Nahecaris balssi,* like its larger relative *Nahecaris stuertzi,* was presumably a good swimmer.

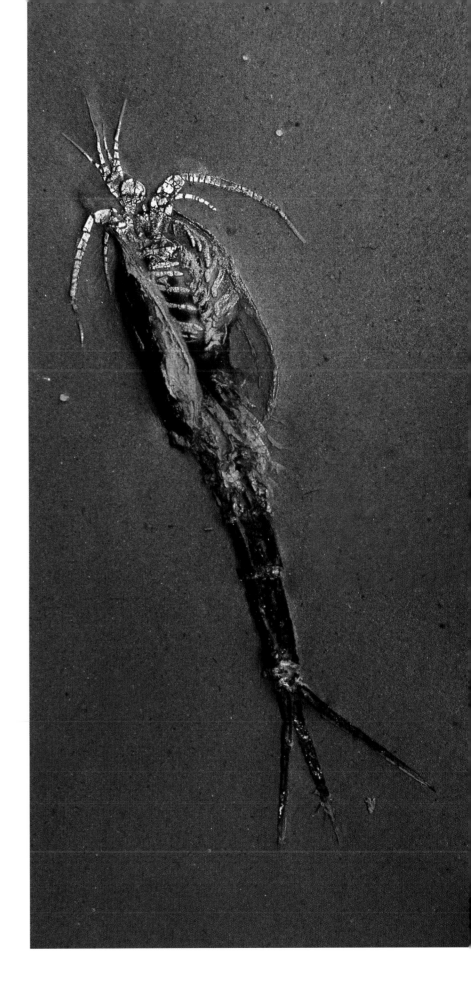

Figure 105. A specimen of *Nahecaris balssi* exposed from the ventral side (length about 6 centimeters).

Fishes (Pisces)

No modern marine fauna would be complete without fish, and a great variety was already present in the Devonian. The placoderms, a very striking group of early jawed fish that reached lengths of up to 10 meters, are the most diverse in the Hunsrück Slate, where some ten species are known, but usually only as fragments. They are mostly flat forms that presumably lived on the seabed. A bony armor gave them their distinctive appearance. It provided protection from predators; placoderms had little to fear, even from sharks.

Gemuendina stuertzi

Gemuendina stuertzi is the most completely known placoderm found in the Hunsrück Slate. The body, including the fins, is covered in small scales, which explains why the outline of this placoderm is often preserved (Figure 106). Only the head is protected by bony plates. The striking appearance of *Gemuendina* gives it an iconic status among Hunsrück Slate fossils (Figure 107).

Specimens of *Gemuendina stuertzi* vary greatly in length, and it is not known at what size it reached sexual maturity. The largest known specimen is about 1 meter long.

Drepanaspis gemuendensis

The armored fish *Drepanaspis,* which had a kind of sucking mouth, belongs to a group of jawless fish called heterostracans (Figure 108). *Drepanaspis* is represented by a number of complete specimens found at Gemünden in the central Hunsrück region. Disarticulated remains have been discovered at other sites in the Taunus Mountains and in the mines of the Lindenschied area of the Hunsrück region, but *Drepanaspis* is not present at Bundenbach where most of the fossils portrayed in this book were found. Examples are known that reached lengths up to 60 centimeters. The specimen pictured here, which is unusually complete, is a juvenile only 16 centimeters long.

The body of *Drepanaspis* was covered with bony plates. The large dorsal and ventral plates and the elongate, tapering lateral plates, which covered the gills of this flat, flounder-shaped fish, are particularly characteristic. Smaller, scalelike elements covered with tubercles filled in the areas between the larger plates, and the trunk and tail were covered with robust scales

Figure 106. Detail of the left fin of *Gemuendina stuertzi* showing the scales (area illustrated about 4 centimeters wide).

facilitating movement. The mouth opening was at the lower front edge of the broad head. *Drepanaspis* probably grazed on microbial films and possibly algal mats on the seafloor. The direction of transport and nature of the sediment at Gemünden indicate that this locality lay south of offshore islands in the Hunsrück sea. Salinity may have been reduced in this area due to water coming off the land. The striking concentration of specimens of *Drepanaspis* at Gemünden suggests that it and other armored Devonian fish preferred such brackish water.

Figure 107. Placoderm *Gemuendina stuertzi* (length 24 centimeters). Part of the right fin and the rear part of the tail have been restored.

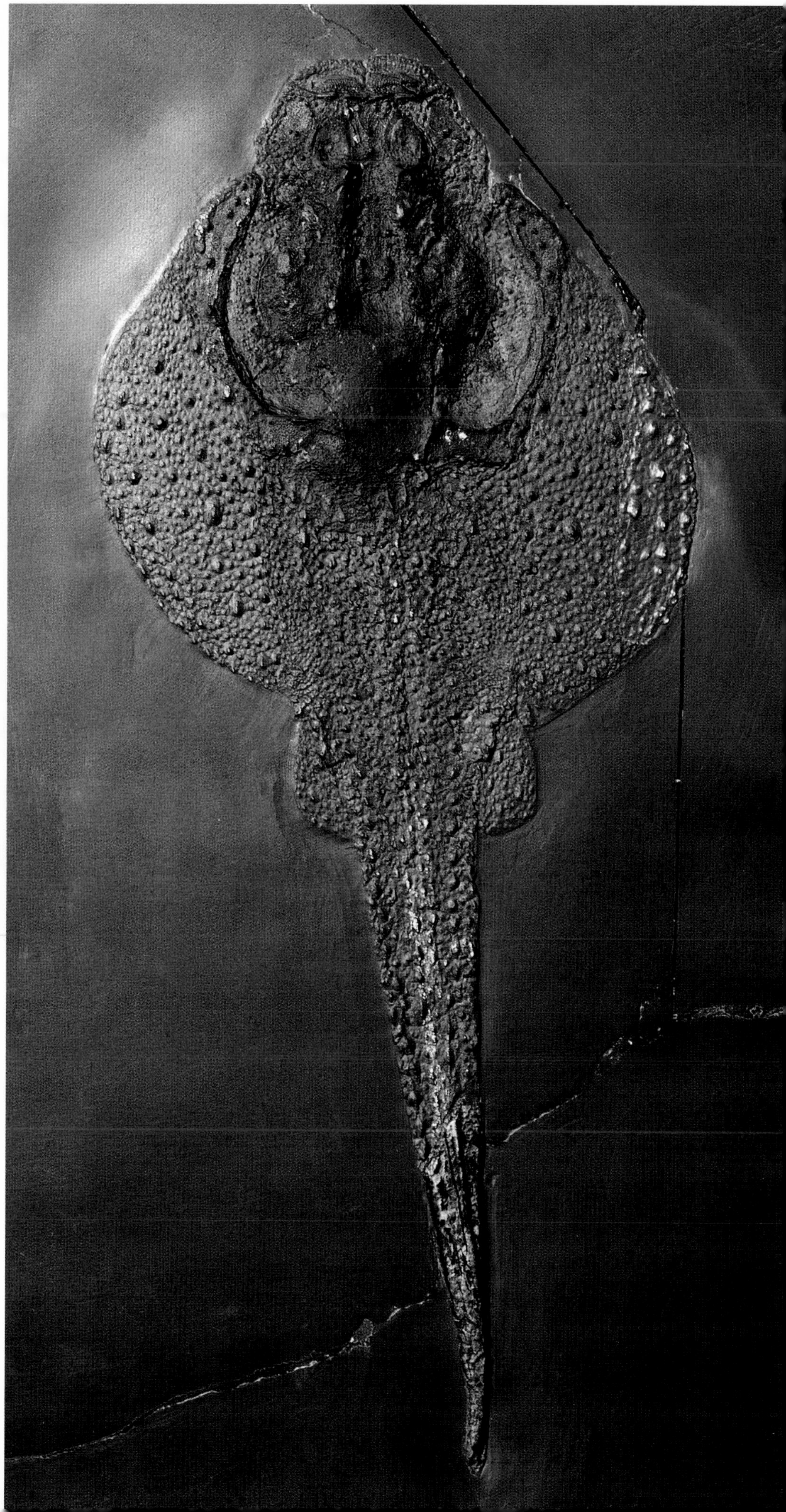

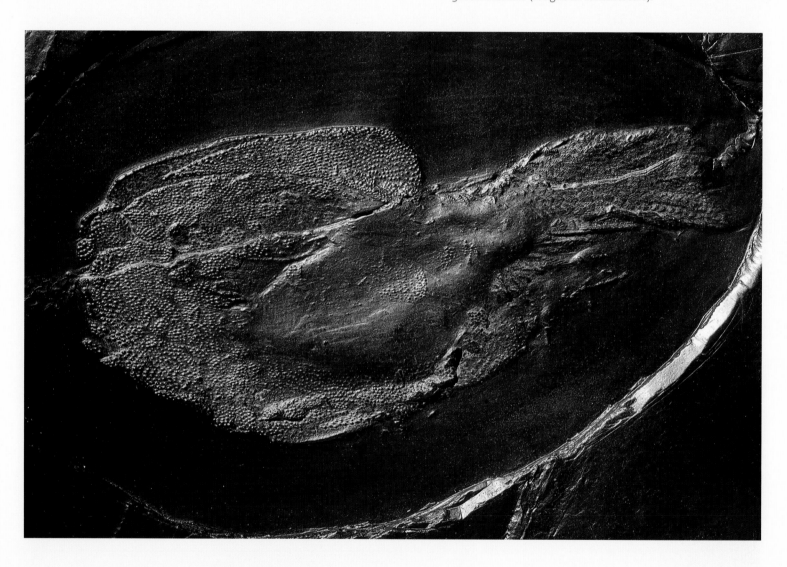

Figure 108. High-quality cast of a specimen of *Drepanaspis gemuendensis* (length 16 centimeters).

Spiny Sharks (Acanthodii)

Also represented in the Hunsrück Slate is a third group of fish, the acanthodians, or spiny sharks (Figure 109). Acanthodians were probably similar in appearance to living sharks, although they are not true sharks and are generally considered an early extinct offshoot of the bony fishes. The fins of spiny sharks were strengthened at the front edge by a large spine. Unfortunately, these spines are often the only feature that is fossilized (Figures 110 and 111). In addition, they are very rare in the Hunsrück Slate. Acanthodian spines are often found in rocks of equivalent age in the Eifel region, however, where they show a high degree of variability. There was clearly a considerable diversity of spiny sharks in the Devonian seas. The spine from the Hunsrück Slate illustrated here is very massive, and it proved possible to expose it from both sides during preparation.

Figure 109. Reconstruction of a spiny shark.

Figure 110. Spiny shark spine and part of associated shoulder element preserved in three dimensions— front (length 22 centimeters).

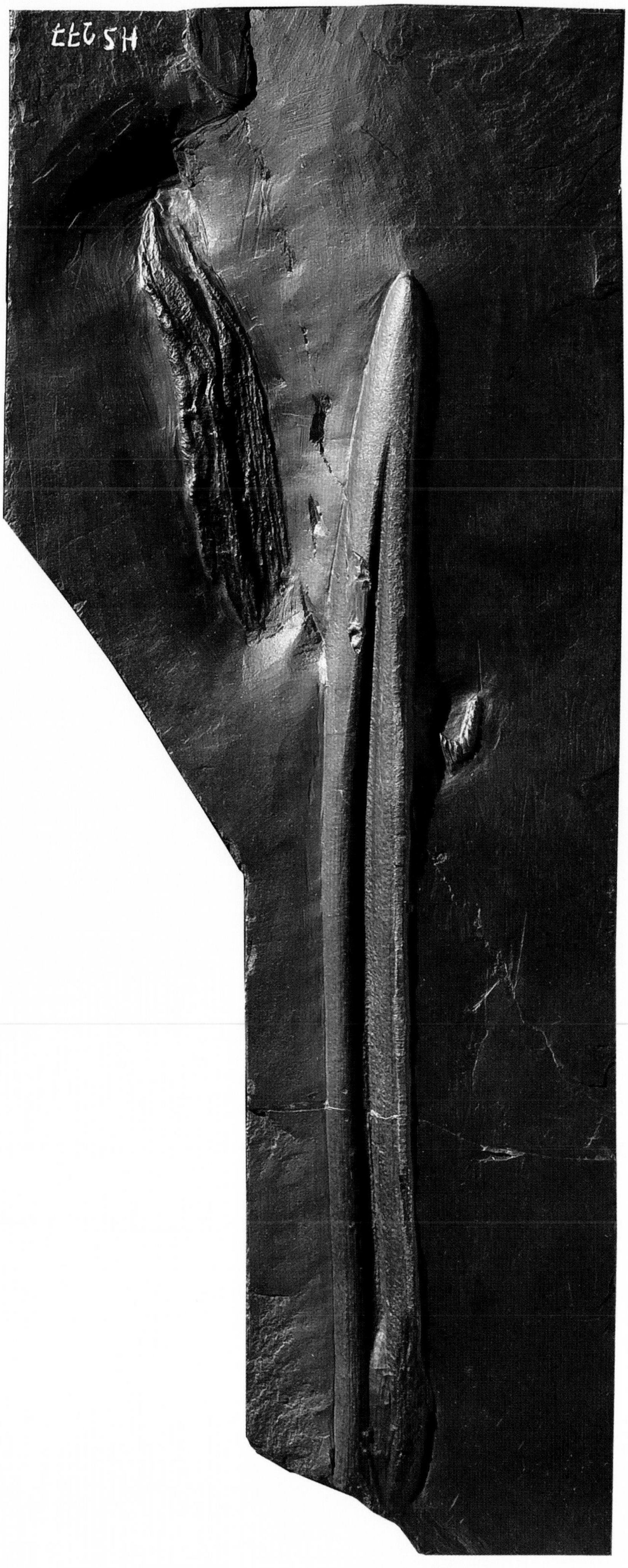

Figure 111. Spiny shark spine and part of associated shoulder element preserved in three dimensions—rear (length 22 centimeters).

OTHER KINDS OF
FOSSIL EVIDENCE

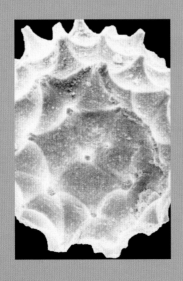  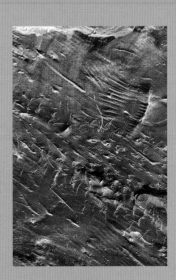

MICROSCOPIC WONDER WORLD

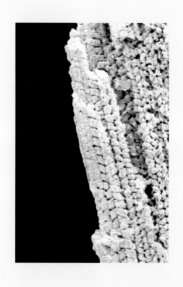  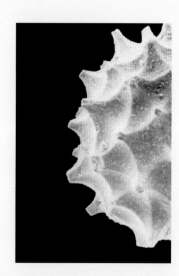

Plant Cells, Spores, and Acritarchs

Most of the fossils from the Hunsrück Slate are marine animals, and some of them are also found elsewhere in sandier sediments deposited close to the coastline. The presence of a few plant fossils in the Hunsrück Slate means that land was nearby. Some areas in the central Hunsrück basin probably formed small islands, such as in the vicinity of Gemünden, where the sediments and their fossils give evidence of temporary brackish water conditions. The Old Red Continent lay to the north, running roughly from Essen to Aachen in Germany and Namur in Belgium. Currents carried plant remains out to sea, where they sank to the seafloor and were buried and then fossilized.

But it is not only large fossils that provide evidence of environments—microfossils are also important. Pyritized plant cells, spores, and acritarchs, some of them almost perfectly preserved, were described from the Hunsrück Slate in 2003.[31] They include bundles of spiral-shaped, water-conducting cells of an unknown land plant (Figure 112). The cells are hexagonal when viewed end-on, in cross-section. Organic material rarely survives. The walls of the cells are usually pyritized and can be distinguished from the coarser pyrite that infills them.

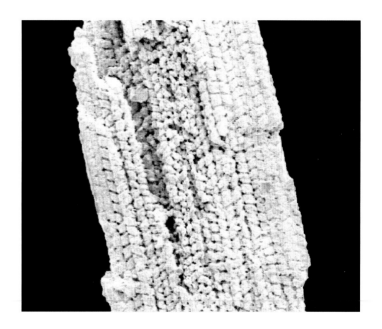

Figure 112. Detail of water-conducting plant cells (width of fossil about 0.5 millimeter).

A three-dimensionally preserved spore with three slitlike openings in the form of a Y, where the spore opens, is also shown here (Figure 113). The outer organic wall of this 400-million-year-old spore has survived, but the interior is completely filled with pyrite. Like the water-conducting cells, it provides evidence of land plants.

Acritarchs are organic walled microfossils (Figure 114). They were part of the marine plankton, so their presence in the Hunsrück Slate is no surprise. It is not yet clear, however, what they are. Acritarchs have been interpreted as the reproductive cysts of algae, but at least some of them may represent the egg cases of arthropods or other animals. The organic walls of acritarchs were resistant to decay, and the wall of the specimen shown here survived long enough to allow the interior to be filled by pyrite before the wall degraded. Acritarchs are very common in Paleozoic rocks and extend back over 3 billion years into the Archaean. They are sometimes useful for correlating sedimentary sequences.

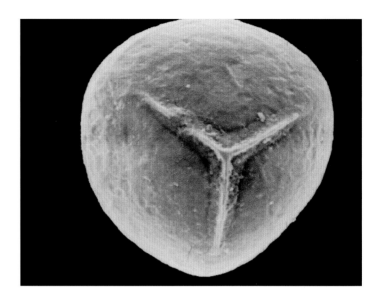

Figure 113. Three-dimensionally preserved spore (diameter about 0.15 millimeter).

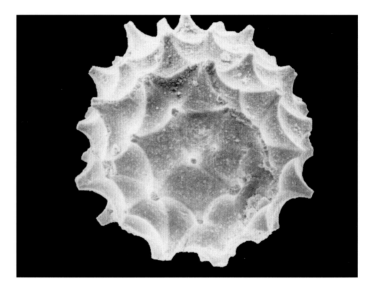

Figure 114. Three-dimensionally preserved acritarch from the Hunsrück Slate (diameter about 0.15 millimeter).

STORIES IN SLATE

Unhappy Endings

We met the mitrate *Rhenocystis latipedunculata* earlier in this book (see Figure 29). *Rhenocystis* moved along the surface of the seafloor with the help of the tail. Two such individuals suffered an unfortunate fate as they tried to escape the deposits of a turbidity current. One was probably already so weakened that its attempt to escape left just a feeble circular impression (Figure 115). The other left a straighter trace, before dying at the end (Figure 116). Both traces show that *Rhenocystis* moved tail first. Death trails are known from other famous fossil deposits, including spiral traces produced by the horseshoe crab *Mesolimulus* in the famous Jurassic Solnhofen Limestone in Bavaria. In that case, however, the horseshoe crab died on the

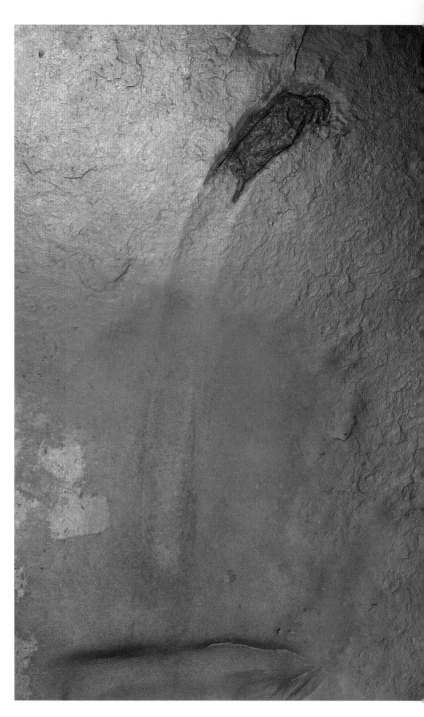

Figure 116. *Rhenocystis latipedunculata* with a long escape trace (area illustrated 10 by 15 centimeters).

Figure 115. *Rhenocystis latipedunculata*, preserved after moving clockwise through a tight circle leaving a faint trace on the surface of the slate (diameter of trace 6 centimeters).

sediment surface, probably due to low oxygen resulting from a lack of currents on the seafloor. Freshwater snails show similar behavior today, dying when the oxygen content drops dramatically in small ponds. *Rhenocystis,* on the other hand, was trying to escape upward after being buried by an influx of sediment.

Lepidocoleus hohensteini is a type of extinct armored bristle worm (polychaete annelid) known as a machaeridian. The example shown is one of fewer than ten known specimens of machaeridians from the Hunsrück Slate, all of which are examples of the same species. They reached lengths of almost 3 centimeters. The body of *Lepidocoleus* was enclosed by plates, which provided protection. Just one of the specimens preserves evidence of the soft parts. An x-radiograph showed that the appendages with bristles used for locomotion by other polychaetes (as on pp. 57–58) were absent.[32] *Lepidocoleus* burrowed through the sediment by passing waves of expansion and contraction along the body and using the plates to gain purchase on the mud. The individual found here, in contrast, died at the end of an unusual meandering S-shaped trace, presumably trying to escape upward through unconsolidated mud in pursuit of more oxygenated conditions (Figure 117).

Figure 117. *Lepidocoleus hohensteini* at the end of an S-shaped trace preserved in the slate (length of worm 3 centimeters).

Eat or Be Eaten

The fossils provide information beyond the nature of the animals in the Hunsrück Slate sea and aspects of their mode of life. Some specimens preserve evidence of the dangers of life during the Devonian. Even large and armored organisms had serious enemies. Complete trilobites are known with indentations on the cephalon (Figure 118). This head shield protected the trilobite's internal organs, including the stomach and the anterior muscles. The culprit in question may have been a placoderm fish. These predators, which reached large sizes, did not have true teeth, but reinforced elements in the jaw bore projections suitable for biting and crushing.

A number of specimens of sea stars and brittle stars have been found with part of an arm lying to one side, as if it had been cut off during life. These fragments are preserved with spines attached and skeletal plates articulated, and they are not separated by any distance from the rest of the animal (Figures 119 and 120). Many echinoderms have the ability to shed part of or even an entire arm in times of stress (autotomy). Being swept up and buried in a cloud of sediment might well have triggered this behavior.

The skeletal elements of crinoids can sometimes be identified in coprolites (fossil feces). Snail shells are more unusual, but an unknown predator was very busy at least occasionally hunting them down for food. The shells were largely undamaged as they passed through the gut. A narrow strip of one such fossil contains more than twenty-five closely spaced snail shells on the surface of a slate slab, all neatly lined up (Figure 121). The tips of the narrow cone-shaped shells tend to face in the same direction, suggesting that the predator preferred to swallow them in a particular way.

Figure 118. A specimen of the trilobite *Chotecops* sp. with possible bite marks on the head shield (length 9 centimeters).

[116]

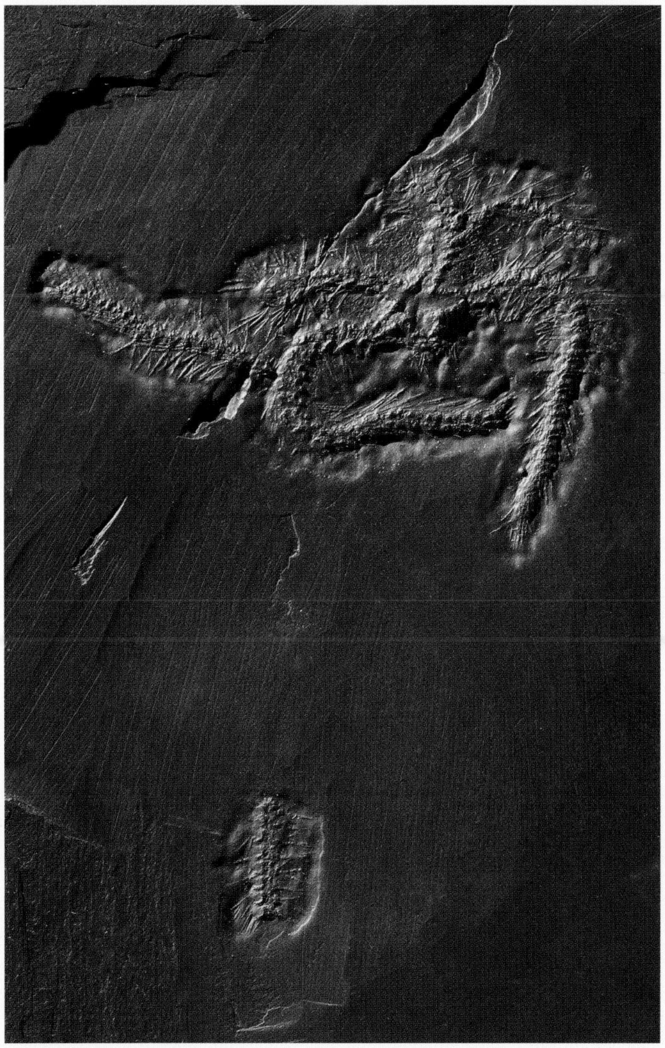

Figure 119. Specimen of the brittle star *Eospondylus primigenius* with severed arm (maximum dimension of the specimen 6 centimeters).

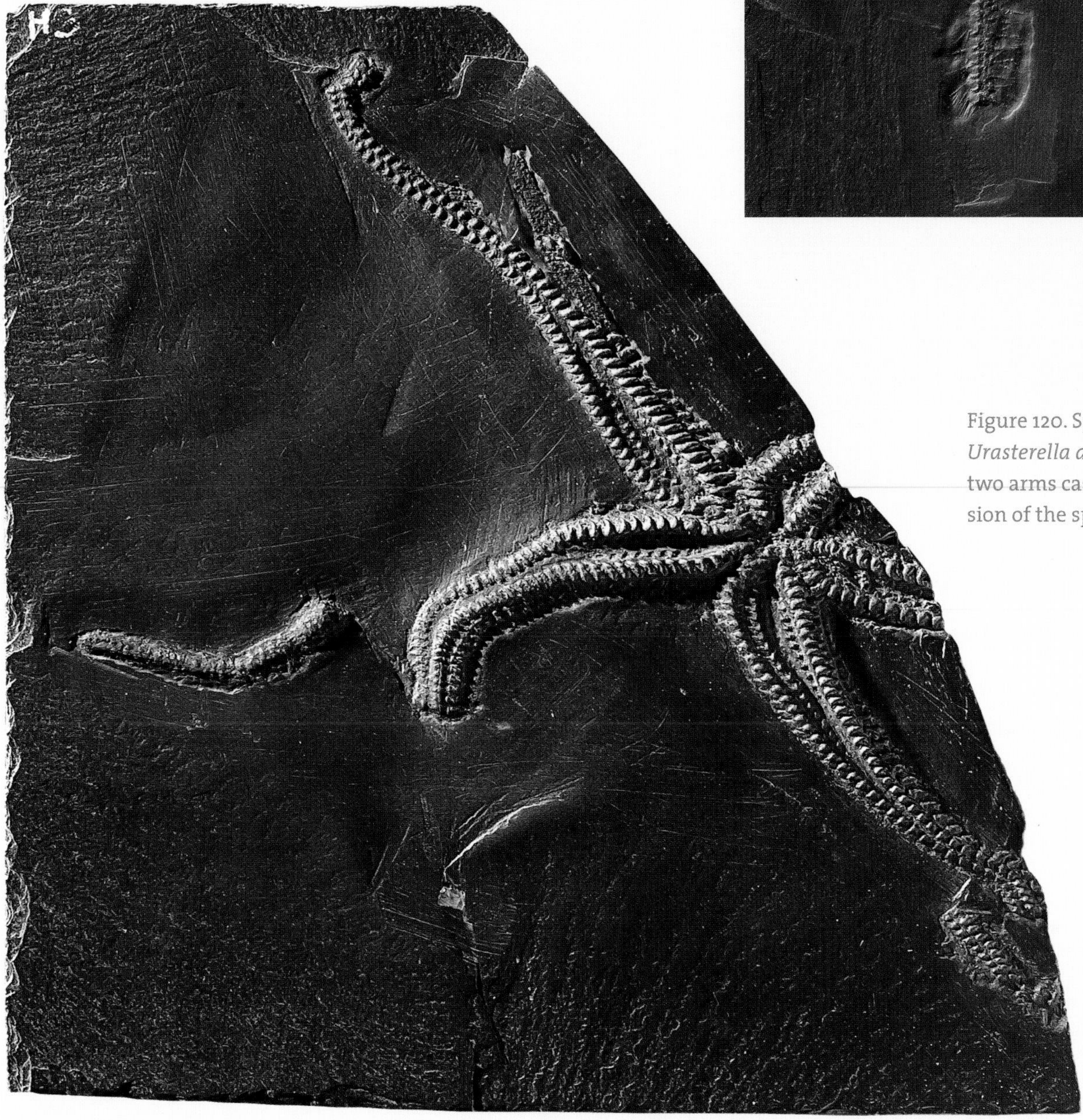

Figure 120. Specimen of the sea star *Urasterella asperula* with the ends of two arms cast off (maximum dimension of the specimen 11 centimeters).

Figure 121. A coprolite from an unknown consumer of snails (length of the specimen 20 centimeters).

Coprolite Concentrations

Roof slate is still produced in a mine in the northern region of the Hunsrück Mountains, in the area around Altlay near the Moselle River. Occasional fossils have been found, but this area lacks the abundance, diversity, and excellent preservation characteristic of the central Hunsrück region. An unusual feature, however, is the coprolites of larger animals. Rapid burial in fine-grained sediment followed by initial decay resulted in conditions that promoted the replacement of feces by pyrite. Some beds yield concentrations, probably produced by fish. These coprolites, both large and small, and sometimes neatly coiled, represent a golden legacy of excretion (Figure 122).

In addition to examples that consist of numerous pill-like spheres, prompting visions of some kind of marine "rabbit" in the Hunsrück sea, there are more interesting types of coprolite that suggest the size of the producer. They are usually preserved in three dimensions, which means there was a rapid replication in pyrite that ultimately proved robust enough to resist tectonic deformation.

The Altlay mine has also yielded a remarkable specimen containing an unusually high proportion of indigestible shelly material (Figure 123). Apart from an abundance of complete and fragmentary brachiopods and bivalves, there are elements of the stems, arms, and cup of crinoids, tentaculitids, and unidentifiable pieces of shell. These residues show some sign of degradation, and occupy an area of about 20 by 5 centimeters. This example, like other fossil coprolites, provided a suitable environment for the recrystallization of minerals in the slate under the influence of temperature and pressure. This process involved remobilization of iron sulfide and the growth of larger pyrite crystals. Aggregates of large and small glittering pyrite crystals delineate this large coprolite, like little jewels in one of nature's more eccentric art forms.

Figure 122. A layer 2 centimeters thick containing numerous pill-shaped coprolites, probably from fish.

Figure 123. Coprolite of an unknown animal (dimensions 20 by 5 centimeters).

Tracks in the Mud

Turbulent Times, Still Water

Trace fossils usually tell a story. Only rarely is the maker of the trace known, because they are hardly ever found together (see Figures 115, 116, and 117). Nevertheless, traces provide evidence of the activities of an animal on the seafloor. The arthropod *Cheloniellon calmani* was a predator that lived mainly on the sediment surface (Figure 124). One trail about 6 centimeters wide consists of small ridges and depressions preserved on a slate slab (Figure 126). Fine lines reveal that the trail was made by an animal with filaments on its limbs. The width of the track, and analysis of the arrangement of the footprints, suggest that this trail may have been the work of *Cheloniellon* moving across the mud.

Another trail may have been made by the horseshoe crab *Weinbergina opitzi* (Figure 125).[33] Two different arthropods left their traces on the same slab (Figure 127). One represents a leisurely walk on the seafloor, while the other animal only touches the mud surface while being carried along by a current. The fine, linear, sometimes hook-shaped tracks may have been made by *Weinbergina*. It probably moved from the top left to the bottom right of the slate slab. Flanking this trackway are comblike impressions. At one point these impressions are traversed and truncated by a deeper groove. These prints were probably made by a trilobite, but it was not running across the seafloor. The traces were produced by the limbs and occasionally by the edge of the exoskeleton. The animal was probably transported by a strong current but made contact with the muddy substrate every few centimeters.

Both trackways cannot have been made at the same time, as they record very different conditions. Probably the trilobite was first transported by a turbidity current, causing the comblike impressions as it made contact with the seabed. Afterward, fine sediment was deposited from the current, covering the marks. *Weinbergina*, or perhaps a second trilobite, walked across the surface leaving tracks that penetrated the soft mud, before it was again covered by sediment during the next depositional event. Both trackways were exposed when the slate was split.

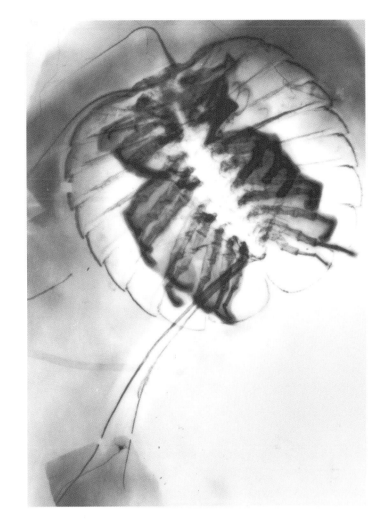

Figure 124. X-radiograph of the Hunsrück Slate arthropod *Cheloniellon calmani* (length 10.5 centimeters).

Figure 125. Reconstruction of the arthropod *Weinbergina opitzi,* after Stürmer and Bergström, "*Weinbergina,* a Xiphosuran Arthropod" (1981).

Figure 126. A trace fossil that may have been made by the arthropod *Cheloniellon calmani* (slate slab 22 centimeters long).

Figure 127. Detail of trace fossils known as *Dimorphichnus* (produced by *Weinbergina*) and *Monomorphichnus* (produced by a trilobite). The area illustrated is 7 by 10 centimeters.

Repositories of Illustrated Specimens

Abbreviations

MNHM: Naturhistorisches Museum Mainz/Landessammlung für Naturkunde Rheinland-Pfalz

DBM: Deutsches Bergbau-Museum (German Mining Museum), Bochum

STIBP: Goldfuß Museum of the Steinmann Institute of the University of Bonn

KGM: Schloßpark Museum, Bad Kreuznach

WS and WB indicate that x-radiographs are the work of Wilhelm Stürmer and Wolfram Blind, respectively.

Figure 9: MNHM PWL1995/15-LS; 10: DBM HS747; 11: MNHM PWL2000/91-LS; 12: MNHM PWL2008/147-LS; 13: MNHM PWL2002/204-LS; 14: STIBP-C1; 15: DBM HS19; 16: MNHM PWL1959/247-LS; 17: DBM HS31; 18: STIBP-C2; 19: DBM HS612; 20: DBM HS724; 21: MNHM PWL 1993/394-LS; 22: DBM HS34; 23: DBM HS72; 24: DBM HS926b; 25: MNHM PWL2006/24-LS (x-radiograph WB703); 26: MNHM PWL2006/24-LS; 28: DBM HS297; 29: DBM HS345; 30: DBM HS323 (x-radiograph WS12854); 31: DBM HS323; 32: DBM HS353; 33: DBM HS353 (x-radiograph WB170); 34: DBM HS353; 35: MNHM PWL2009/71-LS; 36: DBM HS934; 37: MNHM PWL2009/57-LS; 38: MNHM PWL2009/57-LS; 39: DBM HS509; 40: DBM HS572; 41: DBM HS572; 42: DBM HS104; 43: DBM HS104; 45: DBM HS119; 46: DBM HS119; 47: DBM HS133; 48: HS926; 49: DBM HS799; 50: DBM HS208; 51: DBM HS926b; 52: MNHM PWL2007/8-LS; 53: DBM HS768; 54: MNHM PWL2009/2-LS; 55: MNHM PWL1998/131-LS; 56: DBM HS 717 (x-radiograph WB296); 57: MNHM PWL1993/353-LS; 58: MNHM PWL1993/353-LS; 59: MNHM PWL1993/353-LS; 60: MNHM PWL1994/53-LS; 61: STIPB Oltmann 1; 62: (top) MNHM PWL2006/24-LS; (middle) KGM 1987/5041; (below) DBM HS905; 63: DBM HS718; 64: DBM HS867; 65: DBM HS850; 66: DBM HS867; 67: DBM HS580; 68: DBM HS265; 69: MNHM PWL2005/11-LS; 70: DBM HS552; 71: MNHM PWL2007/19-LS; 72: MNHM PWL2007/19-LS; 74: DBM HS582; 75: DBM HS206; 76: MNHM PWL1997/3-LS (x-radiograph WB600); 77: DBM HS285; 81: DBM HS862; 82: STIBP Eb0408; 83: DBM HS512; 84: DBM HS823; 85: STIBP Egr229; 86: STIBP OW242; 87: DBM HS330; 89: MNHM PWL2002/228-LS; 90: MNHM PWL2007/8-LS; 91: DBM WB588; 93: DBM HS576 (x-radiograph WB92); 94: DBM HS602 (x-radiograph WB1); 95: DBM HS824; 96: DBM HS577 (x-radiograph WB86); 97: DBM HS371; 98: DBM HS371; 100: MNHM PWL1994/52-LS; 102: STIBP HuB04Bx; 103: MNHM PWL2009/70-LS; 104: MNHM PWL1993/245-LS; 105: MNHM PWL1999/2-LS; 106: DBM HS642; 107: DBM HS642; 108: DBM WS3949; 110: DBM HS277; 111: DBM HS277; 112: MNHM PB2001/5769b-LS; 113: MNHM PB2001/5775e-LS; 114: MNHM PB2001/5771h-LS; 115: DBM HS524; 116: DBM HS834; 117: DBM HS735; 118: DBM HS723; 119: DBM HS935; 120: DBM HS600; 121: DBM HS936; 122: DBM HS933; 123: DBM HS913; 124: KGM 1983/269 (x-radiograph WS2487); 126: DBM HSM5; 127: DBM HSM62.

Photo Credits

Christoph Bartels
Deutsches Bergbau-Museum, Bochum
(figure number) 7, 8, 26

Alexandra Bergmann
Steinmann Institute, Bonn
5, 32, 38, 48, 50, 56, 57, 62 (middle), 63–65, 72, 98, 101, 105, 106, 118

Wolfram Blind
25, 33, 56, 76 (right), 93, 94, 96

Derek E. G. Briggs
Yale University, New Haven, Connecticut
4

Elke Gröning
Clausthal University of Technology, Germany
99

Peter Hohenstein
Lautertal, Germany
90

Otto Jaekel
44

Gabriele Kühl
Steinmann Institute, Bonn
1, 2, 3 (after Stets & Schäfer 2002), 88, 109, 125 (after Stürmer & Bergström 1978)

Sarah Long (née Tibbs)
Ulverston, United Kingdom
112–114

Carsten Lüter
Museum für Naturkunde, Berlin
27

Georg Oleschinski
Steinmann Institute, Bonn
9–15, 17–20, 22–24, 28, 29, 31, 34–37, 39–43, 45–47, 49, 51–55, 58–62, 65 (top), 66–71, 73–75, 77–87, 89, 91, 95, 97, 100, 102–104, 107, 108, 110, 111, 115–117, 119–123, 126, 127

Astrid Opel
Deutsches Bergbau-Museum, Bochum
16, 21

Simon Powell
University of Bristol
76 (left)

Ferdinand Roemer
92

Wilhelm Stürmer
30, 124

Michael Wuttke
Generaldirektion Kulturelles Erbe RLP, Mainz
6

Further Reading

C. Bartels, D. E. G. Briggs, and G. Brassel, *The Fossils of the Hunsrück Slate: Marine Life in the Devonian* (Cambridge: Cambridge University Press, 1998), 232 pp.

D. J. Bottjer, W. Etter, J. W. Hagadorn, and C. M. Tang, eds., *Exceptional Fossil Preservation: A Unique View on the Evolution of Marine Life* (New York: Columbia University Press, 2002), 403 pp.

P. Selden and J. R. Nudds, *Evolution of Fossil Ecosystems* (Chicago: University of Chicago Press, 2005), 192 pp.

Notes

1. J. Stets and A. Schäfer, "Depositional Environments in the Lower Devonian Siliciclastics of the Rhenohercynian Basin (Rheinisches Schiefergebirge, W-Germany): Case Studies and a Model," *Contributions to Sedimentary Geology* 22 (2002): 78 pp. (Stuttgart: Schweizerbart).

2. D. E. G. Briggs, R. Raiswell, S. H. Bottrell, D. Hatfield, and C. Bartels, "Controls on the Pyritization of Exceptionally Preserved Fossils: An Analysis of the Lower Devonian Hunsrück Slate of Germany," *American Journal of Science* 296 (1996): 633–663.

3. S. T. Grimes, F. Brock, D. Rickard, K. L. Davies, D. Edwards, D. E. G. Briggs, and R. J. Parkes, "Understanding Fossilization: Experimental Pyritization of Plants," *Geology* 29 (2001): 123–126.

4. C. F. Roemer, "Neue Asteriden und Crinoiden aus Devonischem Dachschiefer von Bundenbach und Birkenfeld," *Palaeontographica* 9 (1863): 143–152.

5. O. Follman, "Unterdevonische Crinoiden," *Verhandlungen des naturhistorischen Vereines der preussischen Rheinlande und Westfalens (Bonn)* 44 (1887): 117–173.

6. O. Jaekel, "Beiträge zur Kenntnis der palaeozoischen Crinoiden Deutschlands," *Palaeontologische Abhandlungen,* Neue Folge Band III, Heft I (1895): 1–176 (Jena: Gustav Fischer).

7. A. Seilacher and C. Hemleben, "Spurenfauna und Bildungstiefe der Hunsrückschiefer (Unterdevon)," *Notizblatt des hessischen Landesamtes für Bodenforschung zu Wiesbaden* 94 (1966): 40–53.

8. F. Kutscher, "Zur Entstehung des Hunsrückschiefers am Mittelrhein und auf dem Hunsrück," *Jahrbuch des Nassauischen Vereins für Naturkunde* 81 (1931): 177–232; R. Richter, "Tierwelt und Umwelt im Hunsrückschiefer; zur Enstehung eines schwarzen Schlammsteins," *Senckenbergiana, Frankfurt* 13 (1931): 299–342.

9. C. Bartels and G. Brassel, *Fossilien im Hunsrückschiefer: Dokumente des Meereslebens im Devon* (Museum Idar-Oberstein, 1990), 232 pp.

10. C. Bartels, M. Wuttke, and D. E. G. Briggs, eds., "The *Nahecaris* Project: Releasing the Marine Life of the Devonian from the Hunsrück Slate of Bundenbach," *Metalla (Bochum)* 9 (2002): 55–138.

11. Briggs et al., "Controls on the Pyritization of Exceptionally Preserved Fossils."

12. W. M. Lehmann, "Röntgenuntersuchung von *Asteropyge* sp. Broili aus dem rheinischen Unterdevon," *Neues Jahrbuch für Mineralogie, Geologie und Paläontologie* 72B (1934): 1–14.

13. W. Stürmer, "Soft Parts of Cephalopods and Trilobites: Some Surprising Results of X-ray Examination," *Science* 170 (1970): 1300–1302.

14. Bartels, Wuttke, and Briggs, eds., "The *Nahecaris* Project."

15. C. Bartels and M. Poschmann, "Linguloid Brachiopods with Preserved Pedicles: Occurrence and Taphonomy (Hunsrück Slate, Lower Emsian, Hunsrück, SW Germany)," *Metalla (Bochum)* 9 (2002): 123–130.

16. W. M. Lehmann, "*Pentremitella osoleae* n.g. n.sp., ein Blastoid aus dem unter-devonischen Hunsrückschiefer," *Neues Jahrbuch für Mineralogie, Geologie und Paläontologie Monatshefte* (1949): 186–191.

17. K. Fauchald and E. L. Yochelson, "A Tubicolous Animal from the Hunsrück Slate, West Germany," *Paläontologische Zeitschrift* 64 (1990): 15–23.

18. Jaekel, "Beiträge zur Kenntnis der palaeozoischen Crinoiden Deutschlands."

19. C. Bartels and W. Blind, "Röntgenuntersuchung pyritisch vererzter Fossilien aus dem Hunsrückschiefer (Unter-Devon, Rheinisches Schiefergebirge)," *Metalla (Bochum)* 12 (1995): 79–100.

20. G. Kühl and J. Rust, "Re-investigation of *Mimetaster hexagonalis*: A Marrellomorph Arthropod from the Lower Devonian Hunsrück Slate (Germany)," *Paläontologische Zeitschrift* 84 (2010): 397–411.

21. G. Kühl, J. Bergström, and J. Rust, "Morphology, Palaeobiology, and Phylogenetic Position of *Vachonisia rogeri* (Arthropoda) from the Devonian Hunsrück Slate (Germany)," *Palaeontographica* (A) 286 (2008): 123–157.

22. F. Broili, "Crustaceenfunde aus dem rheinischen Unterdevon," *Sitzungsberichte der Bayerischen Akademie der Wissenschaften Mathematisch-naturwissenschaftliche Abteilung (München)* (1928): 197–201.

23. F. Broili, "Ein neuer Arthropode aus dem rheinischen Unterdevon," *Sitzungsberichte der Bayerischen Akademie der Wissenschaften Mathematisch-naturwissenschaftliche Abteilung (München)* (1929): 135–142.

24. W. Stürmer and J. Bergström, "The Arthropod *Cheloniellon* from the Devonian Hunsrück Shale," *Paläontologische Zeitschrift* 52 (1978): 57–81.

25. D. E. G. Briggs and C. Bartels, "New Arthropods from the Lower Devonian Hunsrück Slate (Lower Emsian, Rhenish Massif, Western Germany)," *Palaeontology* 44 (2001): 275–303.

26. R. A. Moore, D. E. G. Briggs, and C. Bartels, "The Arthropod *Bundenbachiellus giganteus* from the Lower Devonian Hunsrück Slate, Germany," *Paläontologische Zeitschrift* 82 (2008): 31–39.

27. A. Glass and D. B. Blake, "Preservation of Tube Feet in an Ophiuroid (Echinodermata) from the Lower Devonian Hunsrück Slate of Germany and a Redescription of *Bundenbachia beneckei* and *Palaeophiomyxa grandis*," *Paläontologische Zeitschrift* 78 (2004): 73–95.

28. W. M. Lehmann, "Die Asterozoen in den Dachschiefern des rheinischen Unterdevons," *Abhandlungen des Hessischen Landesamtes für Bodenforschung (Wiesbaden)* 21 (1957): 160 pp.

29. G. Kühl, D. E. G. Briggs, and J. Rust, "A Great-Appendage Arthropod with a Radial Mouth from the Lower Devonian Hunsrück Slate, Germany," *Science* 323 (2009): 771–773.

30. P. Van Roy and D. E. G. Briggs, "A Giant Ordovician Anomalocaridid," *Nature* 473 (2011): 510–513.

31. S. L. Tibbs, D. E. G. Briggs, and K. F. Prössl, "Pyritisation of Plant Microfossils from the Devonian Hunsrück Slate of Germany," *Paläontologische Zeitschrift* 77 (2003): 241–246.

32. A. E. S. Högström, D. E. G. Briggs, and C. Bartels, "A Pyritized Lepidocoleid Machaeridian (Annelida) from the Lower Devonian Hunsrück Slate, Germany," *Proceedings of the Royal Society of London* B 276 (2009): 1981–1986.

33. W. Stürmer and J. Bergström, "*Weinbergina,* a Xiphosuran Arthropod from the Devonian Hunsrück Slate," *Paläontologische Zeitschrift* 55 (1981): 237–255.

Index

About the Authors

Gabriele Kühl

Gabriele Kühl studied geology at the Steinmann Institute at the University of Bonn. Her diploma thesis dealt with the arthropod *Vachonisia rogeri* from the Hunsrück Slate and her Ph.D. thesis was on the morphology and phylogeny of other arthropods from the Hunsrück Slate.

Derek E. G. Briggs

Derek Briggs is a paleontologist at Yale University, New Haven, Connecticut, where he is the G. Evelyn Hutchinson Professor of Geology and Geophysics and director of the Yale Peabody Museum of Natural History. He explored the extraordinary conditions that led to the pyritization of the Hunsrück Slate and described some of the Hunsrück Slate fossils, including arthropods and annelids. He is co-author of the book *The Fossils of the Hunsrück Slate: Marine Life in the Devonian* with Christoph Bartels and Günther Brassel. In 2008, he was a research awardee of the Alexander von Humboldt Foundation at the University of Bonn.

Christoph Bartels

Christoph Bartels is a full-time historian. In addition to his work at the German Mining Museum in Bochum, Bartels has collected, prepared, and identified fossils from the Hunsrück Slate since his school days. Together with Günther Brassel, he published *Fossilien im Hunsrückschiefer: Dokumente des Meereslebens im Devon* in 1990. Subsequently, in 1998, he published *The Fossils of the Hunsrück Slate: Marine Life in the Devonian* with Derek Briggs and Günther Brassel. His recent research includes the documentation of the bristle worms from the Hunsrück Slate.

Jes Rust

Jes Rust is a specialist on invertebrate fossils and professor at the Steinmann Institute at the University of Bonn. The focus of his research on the Hunsrück Slate is the paleobiology and evolution of arthropods and tentaculitids, and the conditions that led to the preservation of soft-bodied fossils. His studies of the fossils from the Hunsrück Slate are funded by the German Research Foundation (DFG).

About the Photographers

Georg Oleschinski

Georg Oleschinski trained in a Cologne studio for advertising photography, and has worked at the Steinmann Institute, in the field of paleontology, since 1983. He is a member of the art network NRW and also works as a freelance artist.

Alexandra Bergmann

Alexandra Bergmann is a trained photographer and holds a degree in biology. Her diploma thesis, which she researched at the Steinmann Institute, University of Bonn, dealt with the crustacean *Nahecaris balssi* from the Hunsrück Slate. She is currently a research fellow at the Steinmann Institute.